Electricity in Locomotion

An Account of its Mechanism its Achievements, and its Prospects

Adam Gowens Whyte

Classical Prints

This edition published in 2022

ISBN : 9789356535947

Contents

PREFACE

In the following pages an attempt is made to give a clear picture of the part which electricity has taken and will continue to take in the development of locomotion.

Some of the aspects of electric traction are highly technical; others are purely financial. It is impossible to understand the achievements and possibilities of electricity in locomotion without a certain amount of discussion of both these points of view; but it is not necessary to go deeply into either in order to catch some of the enthusiasm which inspires the electrical engineer in his efforts to extend electric traction everywhere on road and rail. The hopes of electrical conquest extend, indeed, to locomotion on the sea and in the air as well as on the land. At the root of these hopes there lies a firm faith in the superior economies and flexibility of electricity as a mode of motion.

In the explanations which are given of electric tramways, electric railways, electric automobiles, electric propulsion on ships, and the other phases of electric traction, nothing but the most elementary knowledge of electricity is presupposed. A certain amount of technical description is unavoidable, but I have restricted it as far as possible to essential matters which throw light upon the meaning of the various systems of electric traction and explain the economic and physical reasons for their adoption.

Anyone who glances over the history of electric traction will be struck by the absence of outstanding names. There is no man who occupies the same position in the sphere of electric locomotion as Watt does in the world of steam, or Stephenson in the world of railways. As a pioneer, Dr. Wernher von Siemens perhaps deserves more honour than any other. But the leading ideas embodied in electric traction systems were contributed by engineers who worked in the general field of electrical engineering; and they have been applied and developed by a numerous band of men who have added one brick of experience and ingenuity to another until the imposing structure was made visible to the world.

Nevertheless, I hope the story as told briefly in the following chapters will not be found devoid of human interest. It has the advantage, at any rate, of the attraction which anything pertaining to electricity holds for all sections of the public. This attraction deepens upon closer acquaintance with the mechanism and the history of electricity in action; and if any of the descriptions and forecasts are found to be prejudiced in favour of a single instrument of locomotion, the fault may be considered to rest with the spell which electricity throws upon everyone who is concerned in any way with its applications in the service of man.

I have to acknowledge the kind assistance of Mr. Frank Broadbent, M.I.E.E., in looking over the proofs of this volume.

A. G. W.

21 April 1911

CHAPTER I

THE WHEEL AND THE PUBLIC

One of the greatest of unknown men of genius was the inventor of the wheel. Probably—as in the case of most inventions—he shares the credit with others who prepared the way for him by discovering that heavy weights could be more easily rolled than dragged. But, whatever the origin of the wheel and axle, the combination was so admirable that it remained unchanged in its essential features for centuries and still forms the primary element in locomotion.

Some of the earliest forms of vehicle can be found co-existing with the very latest. In Oporto, for instance, there are electric tramways, but there are also ox wagons which seem to belong to the childhood of the world. The wheels are rigidly fixed to rotating axles (the oldest known arrangement) and the supports of both the front and the back axles are rigidly fixed to the wagon. The result is that the vehicle cannot 'steer' and must be dragged round corners. Some time ago the authorities, realising at last that this dragging was ruinous to the road surfaces, made a regulation that all wagons should have their front axles pivoted. This attempt at improvement caused more agitation than the Revolution itself. The owners of wagons argued—with perfect justice—that the rigid wagon had served for innumerable generations; and they refused, in the face of fines, to make the change. Their resistance was so general and so dogged that the law became a dead letter, and the people reverted with great content to the ancient system which divided the business of local transport between yoked oxen and women who had been trained from girlhood to carry heavy loads upon their heads.

This example of conservatism, though extreme, is characteristic of the attitude of the general public towards innovations in locomotion. Until mechanical power came to be used, there was—for many centuries—nothing which could be described as a radical innovation in transport. Roads were multiplied and improved; some advance was made in the design and construction of carriages; and the organisation of posting and stage-coach services was developed. But little more was done. Compared with these superficial changes, the idea of using steam power on the highway or on a railroad was so drastic a change that it roused tremendous opposition. The railway companies fought this opposition and overcame it, but the use of steam carriages on ordinary roads was postponed until the appearance of the petrol motor encouraged a movement—once more against strong prejudice—for the repeal of the legislation which restricted the use of mechanically-propelled vehicles on the roads. In a similar way horse tramways were violently attacked; and their conversion to electric traction was opposed by a determined minority in every town. More recently, there was a vigorous agitation against the substitution of motor omnibuses for horse omnibuses in London and elsewhere.

To some extent this recurrent opposition was reasonable enough. The new forms of locomotion had dangers of their own; they were generally noisy and sometimes dirty; and occasionally, as in the case of early tramways, they were a nuisance to existing traffic. But it may be noted that electricity claims to provide a means of locomotion not only more rapid and more efficient (in most cases) than any other, but free from many of the drawbacks which gave conservatism an excuse for opposing the introduction of steam and other forms of locomotion.

In the following pages I hope to give a clear account of the achievements of electricity in the field of locomotion and also to indicate some of its more immediate potentialities.

CHAPTER II

EARLY TRAMROADS AND RAILWAYS

It has sometimes been remarked, by unfriendly critics, that tramways are an apology for bad roads. That is to say, if road surfaces were perfect, there would be no need to lay rails in order to allow vehicles to run easily.

Although this view of the case may be no better than a quarter-truth, it is justified to the extent that tramways were, as a matter of fact, the outcome of an attempt to escape from bad road surfaces. In the early days of mining, coals were taken by horsedrawn wagons from the pits to the harbours. The passage and re-passage of heavy vehicles on the same roadway led to the formation of deep ruts; and the first step towards both the tramway and the railway was taken when logs of wood or 'trams' were laid in the ruts to facilitate transport.

The next step was to make the upper surface of the log round and the rims of the wheels hollow, so that they fitted over the rails and kept the wagons on the track. Owing to the upper part of the rails wearing away quickly, thin plates of iron were in some cases nailed to them. This improvement led to the adoption of a cast-iron rail, fastened to wooden sleepers.

The earliest cast-iron railway was laid down before the middle of the eighteenth century, about one hundred years after the first wooden 'tramways.' Half a century later we find the first rail-and-wheel combination as we know it on modern tramways and railways, where the wheel carries an inner flange and runs upon the head of a narrow metal rail. This is the form which experience has proved to be best adapted for safety, speed, and economy in power. The improvements made since the beginning of the nineteenth century have been in matters of detail.

Many miles of colliery tramroads were in existence when—at the beginning of the nineteenth century—the idea of using the steam engine in place of the horse was taken up by engineers. They were concerned at first solely with the carriage of coal; the idea of conveying passengers arose at a later date, after the steam automobile had been tried and abandoned for the time being. George Stephenson, for instance, ran his first locomotives on colliery tramroads; and the first railway—between Stockton and Darlington—was used for passengers merely as an afterthought. It was, in fact, designed to be a tramroad for the use of the public in general transport by horse traction.

The most curious feature of this stage in the evolution of locomotion was that, although Stephenson's locomotives had been at work for several years and although several schemes of iron roads had been projected, very few people had any conception of the development awaiting the locomotive and iron road in combination. They did not even appreciate the proved fact that the locomotive was a more efficient means of transport than the horse. An immense amount of pioneering work had to be done before the impression of a new era could be borne in upon the public mind. These were the days when the *Quarterly Review* backed 'old Father Thames against the Woolwich Railway for any sum' and when a witness before a Parliamentary Committee (on the Liverpool and Manchester Railway Bill, in 1825) thought himself safe in suggesting that a steam locomotive could not start against a gale of wind.

When these prejudices were overcome, many years had to pass before the objections of landowners and citizens were worn down. Railway engineers spent most of their time in a form of diplomatic warfare with opponents to their schemes; huge sums—part of which still lingers in the capital accounts of railway companies—were spent in Parliamentary proceedings over Railway Bills. This barren process had to be repeated when electric traction made its appearance; but happily the electrical fight was not upon quite so extensive a scale, nor was the period of preparation followed by anything comparable to the Railway Mania of 1845, when the public made up for its early contempt of railway enterprise by tumbling over itself to get shares in some of the most crazy schemes which were ever put into shape by unscrupulous company promoters.

The early history of the steam railway is interesting in connection with electrical locomotion for two reasons. It shows that the railroad proper evolved out of the tramroad or 'light railway,' as it would now be called—a type of line which is specially suited to electrical operation. It also includes a controversy between three modes of traction; and this controversy forms a very good introduction to a discussion of the reasons why electricity is so economical in locomotion.

These three modes were (1) stationary engines: (2) locomotives: (3) the device known as the 'atmospheric railway.'

In both the first and third, engine houses were placed close to the line at convenient intervals. In the first, each steam engine operated an endless rope to which the train of carriages was attached. The system is still in use for colliery working and is also employed (in an improved form, of course) for funicular railways. George Stephenson himself employed it to assist locomotives up heavy gradients. In the atmospheric railway the stationary engines were used to exhaust the air from a length of cast-iron piping laid close to the railway. The principle is the same as that of the 'pneumatic tube' which the Post Office uses for sending papers over short distances. The papers are placed in a cylinder which fits the interior of the tube; and when the air is exhausted from the tube in front of the cylinder, the pressure of the air behind it drives the cylinder forward.

Nowadays it is difficult to realise that such a system was seriously proposed for railway work and actually adopted by an engineer of such eminence as Brunel. But in point of fact it was recommended by two Board of Trade experts in 1842 and by a Select Committee appointed in 1845 to consider several Bills for atmospheric railways. It was tried at Dalkey and Croydon, and it was installed under Brunel's supervision on a six-mile line in Devon. The carrier in the tube was connected to the train through a longitudinal slit at the top of the tube. The slit was closed by a leather flap, except when momentarily lifted by the passage of the train. A great deal of ingenuity was exhausted in attempting to make this 'longitudinal valve' efficient, but it was found that heat, moisture, and frost made the leather deteriorate so rapidly as to render it hopelessly ineffective in a short time. After a series of misfortunes the atmospheric railway became a mere curiosity in the history of invention.

Stephenson was right in regarding the atmospheric railway as 'only the fixed engine and ropes over again, in another form.' He was also right in his belief that the steam locomotive was more economical than either of its rivals. But the stationary engine idea had the germ of an even sounder principle than that of the locomotive. Both in electric tramways and electric railways the power is obtained from stationary engines. The main difference between the electric system and the old rope and atmospheric systems lies in the superior economy with which the power is conveyed electrically to the trains. There are other important differences; but the essential point is that both rope traction and pneumatic propulsion wasted so much power between the engine and the train that their other advantages were annulled, and it was found cheaper to put the engine on wheels and make it drag itself as well as the train.

Brunel's reasons for his faith in the atmospheric railway are well worth quoting for the light they throw indirectly upon the advantages of electric traction. He argued that stationary power, if freed from incumbrances such as the friction and dead weight of a rope, was superior to locomotive power, on the following grounds:

(*a*) A given amount of power may be supplied by a stationary engine at a less cost than if supplied by a locomotive.

(*b*) The dead weight of a locomotive forms a large proportion of the whole travelling load, and thus inherently involves a proportionate waste of power—a waste which is enhanced by the steepness of the gradients and the speed of the trains.

Experience has proved the soundness of these principles. There has been a steady improvement in the power and efficiency of locomotives, but progress has reached a point at which further increases in speed and accelerating power (a very important matter) are not attainable without a prohibitive increase in the consumption of coal and a costly strengthening of the railway track to stand the strain of heavier engines pounding along at very high speeds. Electric traction, which is a reversion in part to the stationary engine system, offers a means of escape from the limitations of the locomotive.

There is still some doubt in the minds of railway engineers whether electric traction is really superior to the steam locomotive on the main railway lines, where distances are great and train loads heavy. But the superiority is admitted on suburban lines and also on tramways, where electricity has almost completely supplanted both horse and steam traction. If Brunel had foreseen how economical electricity would be in the transmission of power between engine and train, he would have felt still more confident in his defence of the stationary engine.

CHAPTER III

THE BIRTH OF ELECTRIC TRACTION

The story of electric traction really begins in the laboratory of Faraday. He was the first to produce mechanical rotation by electrical means; and, although he had no practical end in view, his investigations produced the germ of the commercial dynamo and thence of the commercial electric motor.

That germ, however, took about half a century to develop. It is true that in 1837 (about ten years after Faraday's discovery) Robert Davidson experimented with an electric locomotive on the Edinburgh and Glasgow Railway; it is also true that Jacobi, two years later, propelled a boat on the Neva with electric power. But these early attempts were not on a commercial scale. Not only was the motor a crude contrivance, but the method of producing the electric power was hopelessly extravagant.

At that period the 'primary battery'—similar in character to those still used for laboratory purposes, ringing electric bells, and so on—was the best available source of electricity. Such batteries generate current by the chemical consumption of zinc. In order to obtain sufficient power to move a boat, a large number of batteries had to be coupled together. They were expensive in first cost, expensive in the zinc which was their 'fuel'; and they became rapidly exhausted.

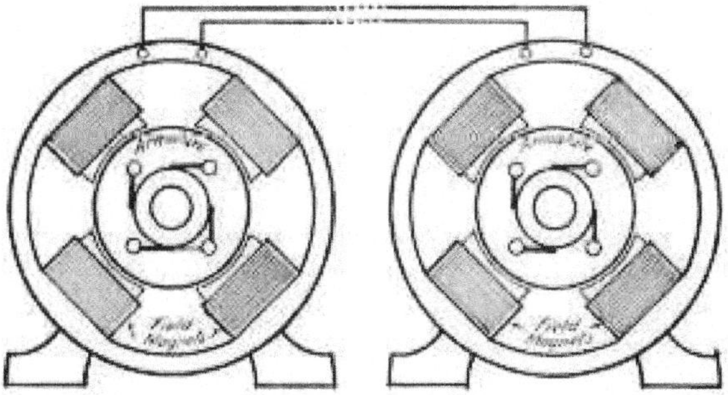

Fig. 1. Diagram to illustrate the essential identity of the dynamo and the motor. The dynamo generates electricity when the armature or group of coils is forcibly revolved close to magnets, thus converting mechanical energy into electrical energy. The motor causes its armature to revolve forcibly when current is supplied to it from the dynamo. Thus the motor converts electrical energy into mechanical energy.

The essential step towards the commercial plane was taken when an efficient means was devised for transforming mechanical into electrical energy on a large scale. The first 'dynamo-electric' machines, invented about the middle of last century, were merely hand machines. Their power was limited by the strength of the permanent magnets employed in their construction; and although an increase in power was obtained by multiplying the number of magnets and driving by steam power, it was not sufficient for commercial purposes. In 1867 electro-magnets were first employed by Siemens and Wheatstone; and from this application there was developed a machine whose power as a generator of electricity was limited only by its size and the speed at which it was run.

It is unnecessary for our present purpose to enter into the technical details of the modern electric generator and the modern electric motor. The principles underlying them are quite simple, although the theory of their design and the practice of their construction and operation are almost a science in themselves. A dynamo or electric generator is a machine for transforming mechanical into electrical energy; an electric motor is a machine for transforming electrical energy into mechanical energy. If, therefore, we place an electric motor upon a vehicle and supply it continuously with current from a dynamo, the motor will rotate and can be used to propel the vehicle. That is the essential mechanism of electric traction.

The simplicity of the arrangement is enhanced by the fact that the dynamo and the motor are virtually the same machine. In the dynamo, a cylindrical 'armature' of coils is forced to rotate close to the poles of electro-magnets; the energy exerted in turning the armature against the influence of the electro-magnets is transformed into the energy of electric currents in the coils of the armature. In the motor, which also consists of an armature close to the poles of electro-magnets, the process is reversed. When a current is passed through the coils of the armature, the reaction between these currents and the electro-magnets causes the armature to revolve.

This reversibility of the dynamo was, according to a story frequently repeated, first discovered quite by accident. In a Paris exhibition a number of Gramme dynamos—or dynamo-electric machines, as they were then called—were being separately connected to lamps and other devices for showing the effect of electric currents; and when one was started up it was found that another was being *driven* at a rapid rate. Investigation showed that the second one had been coupled up to the first by mistake and was therefore being worked as a motor by it.

This was in the year 1879; and the story of the incident served to draw general attention to the discovery of a new and efficient means of transmitting power. Engineers recognised that in the steam-driven dynamo they had the means of producing powerful electric currents, while in the electric motor, connected by wires to the dynamo, they had the means of reproducing the power in mechanical form at a distance. There were, of course, losses of energy in the process. A certain percentage was lost in the dynamo itself, some in the transmitting wires, and some in the motor. But the all-round efficiency of the arrangement was much higher than that of any other system of transmitting power from one point to another several miles distant.

In order to apply this system to propelling vehicles it was only necessary to devise a continuous connection between the motor on the vehicle and the stationary dynamo. This was done on the first electric railway by means of a 'third rail,' substantially in the same way as is now familiar on underground and other electric lines. The third rail was a metal conductor supported on insulators and connected to the dynamo. The vehicle or car was furnished with a metal brush or skate which rubbed along the third rail as the car moved forward. The current thus collected was led through the motor (which drove the axle of the car through toothed wheels) and thence to the track rails, which conveyed the current back to the dynamo and so completed the electrical circuit. Messrs Siemens and Halske exhibited the first electric railway of this type at the Berlin Industrial Exhibition of 1879.

Another method of collecting the current was tried soon afterwards and formed the direct forerunner of the electric tramway on the now standard 'overhead' system. The disadvantage of the third rail system is that it involves an exposed 'live' conductor close to the ground. It is therefore quite unsuited for use on streets. Consequently the next step towards the electric tramway was to carry the electrical conductors overhead by supporting them on poles erected at the side of the track. The first installation of this kind was laid down at the Paris Exhibition of 1881. In that case the conductor was an iron tube with a slot along its lower side; and inside the tube was a 'boat' which slid along and was connected to the car by means of a flexible wire. A second tube, also with a boat and connecting wire, was provided to carry the return current. We shall see later how this arrangement evolved into the familiar 'trolley' system.

The mention of a slotted tube recalls the atmospheric system and, in so doing, emphasises the superiority of the electric system in simplicity, flexibility, reliability, and economy. Brunel's faith in the advantages of stationary engines and the transmission of power therefrom to moving trains would have been justified by the event if the pneumatic system of power transmission had been as practicable as the electric system. But there is an obvious contrast between the huge pipe of the atmospheric railway, with its impossible 'longitudinal valve,' and the small tube of the first overhead electric line or the third rail of the first electric railway. There is also a pathetic contrast between the prolonged struggles which Brunel and the inventors of the atmospheric system underwent before they were forced to acknowledge failure, and the rapid ease with which electric traction entered into its kingdom when the commercial dynamo and motor were first produced. The intrinsic difficulties which electric traction engineers had to meet were not serious. Designers passed, step by step, from the model electric railway at the Berlin Exhibition to public lines on a larger scale, and from the model electric overhead tramway to the 'street railway' or tramway which gradually supplanted the horse tramway. Each step consisted in an extension of the distance covered and an increase in the power required, coincident with a gradual improvement in the details of motors, dynamos, and transmission equipment.

CHAPTER IV

THE ESSENTIAL ADVANTAGES OF ELECTRIC TRACTION ON TRAMWAYS

A railway journal once committed itself to the statement that horse traction was superior to electric traction on roads because the horse possessed the 'vital principle' of energy in its constitution.

It is distinctly curious to find an authority on locomotion describing the essential drawback of horse traction as its distinguishing advantage. The 'vital principle,' unfortunately, needs food and rest to maintain it not only during working hours but during the hours of inactivity as well. In actual practice four horses out of every five in a tramway stud are in the stables while the fifth is at work. Moreover, the same stud has to be kept up, at a practically uniform cost, whether the daily traffic be light or heavy. Thirdly, the 'vital principle' has only a limited number of years during which—apart from sickness and disease—it is effective for traction purposes.

Fig. 2. A typical electric tramway on the overhead system.—The trolley standard carries the wires for supplying current to the cars on both the up and down tracks. The driver has his left hand on the controller handle and his right hand on the brake handle. (Photograph reproduced by courtesy of Dick, Kerr and Company, Limited.)

Another disadvantage is that the pull which a horse can actually exercise on a car is strictly limited and is only a small fraction of the total power represented by the fodder which the horse consumes. The strain upon a horse in starting a car or omnibus is so great that a 'lover of animals' used to supply London omnibuses with appeals to passengers not to stop the omnibus more often than was necessary, especially on an incline. This was a recognition of the fact that the horse cannot cope easily with the heavy strain at starting, and that he requires assistance on heavy gradients.

It was not surprising, therefore, that on horse tramway systems the speed was low, the cars of limited capacity, and the fares comparatively high. The shortness of the journey which a tramway horse was able to cover without fatigue also tended to limit the length of routes.

On all these points electric traction was soon found to be distinctly superior to horse traction. It was more economical in power; it was able to maintain higher speeds with larger and more commodious cars; and there was no narrow limit to the length of routes or the gradients which could be surmounted. Consequently electric traction offered the public an improved service at lower fares.

The whole of the power-producing plant for a typical electric tramway system is concentrated at a generating station placed (if possible) near the centre of the system. From this station runs a network of electric mains to feed the lines with current at convenient points. This concentration is a benefit on several grounds. A large generating equipment is cheaper in first cost than a multitude of small power-producing plants, and it is much more economical in operation. If every car had its own power equipment, that equipment would need to be powerful enough to haul itself and the loaded car up the steepest gradient on the route. That is to say, the sum of the car capacities would be equal to the sum of the maximum demands. But when the power is obtained from a single stationary source we do away with the dead weight of the power equipment on the car, and secure the very vital advantage that the capacity of the stationary source need not be so great as the sum of the maximum demands. In actual working it never happens that all the cars are full of passengers and ascending the steepest gradients simultaneously. While some are running up-hill, others are going down-hill; while some are full, others are half full or almost empty. The result is that the total demand for power at any time is always very much less than the total of the maximum demands made by each car; and the capacity of the generating station need be sufficient to cope only with the smaller amount.

This advantage reduces the expenditure necessary upon boilers, engines, and dynamos at the tramway generating station. And it is enhanced by two valuable capabilities of the electric motor. The first is its power of taking a heavy overload for a limited period without injury. There is no difficulty about making an electric motor, whose normal capacity is 20 horse power, give 40 horse power momentarily, 30 horse power for several minutes, and 25 horse power during the best part of an hour. Applied to tramway work, this advantage means that the rated capacity of the motor equipment of a car may be less than what is required to haul a loaded car at an adequate speed up the steepest gradient on the system. Such maximum demands, which only occur at intervals with each car, can be met by the readiness of the electric motor for overwork. The motors may therefore be reduced in size, saving money in first cost and in the current consumed.

The second valuable peculiarity of the electric motor is that it gives its 'maximum torque' at starting. That is to say, it exercises the highest propulsive effort at the precise moment when it is required. When horses are employed, they have to endure an abnormal strain in overcoming the inertia of a stationary vehicle; everyone must have noticed how horses have to struggle to start a car which they can keep going at an easy trot once it has got up speed. The electric motor—to use an apparent paradox—gives this abnormal pull as part of its normal action. As the inertia of the car is gradually overcome, the speed of rotation of the motor increases and its torque decreases, automatically and precisely in accordance with the demands of the case.

The starting torque of a motor is such an emphatic phenomenon that the driver of an electric car may, if he is careless and switches the current on too suddenly, jerk any standing passenger off his feet, even though the total weight of the car may be ten tons or more. Properly employed, however, the electric motor gives an even and *rapid* acceleration.

This is a far more important point in tramway economics than it appears to be at first sight. The superiority of the electric tramway over the horse tramway depends less upon higher speed than upon the fact that less time is wasted in stopping to pick up and set down passengers. Time is the vital element in all transport, and it is especially vital in connection with tramways, which have to stop with great frequency. If the time which elapses between putting on the brakes at each stop and getting up to full speed again can be materially shortened, then the average speed of the tramway journey can be materially raised. It is easy, by means of powerful brakes, to bring a car to rest quickly; the electric motor enables speed to be regained quickly. In this way a high average speed may be maintained in spite of numerous stops; and, with larger cars, the electric tramway is able to handle a larger volume of traffic in a shorter space of time than the horse tramway.

The time lost in stopping is of so much consequence that, when electric tramways were introduced, the old custom of stopping the cars at any desired point was abandoned. Stopping places were arranged at convenient points along the route, some of them being regular stops and others optional at a signal from passengers desiring to alight or to board the car. The public soon got used to walking a short distance to a stopping place, although they did not, perhaps, appreciate how much the collection of traffic at a reduced number of points tended to improve the general tramway service.

A high average speed with numerous stops was, however, only one of the improvements which the public derived from electric traction. Tramway passengers expect to find a car not only at a convenient point but within a convenient period of waiting. With electric traction the service became much more frequent than with horse traction. It is quite possible to run a horse tramway service profitably with cars at intervals of fifteen to thirty minutes, if the passengers are patient enough to wait and fill each vehicle. But with electric traction the main item is the cost of the standing equipment—the power house, mains, and overhead lines—and unless that equipment is adequately utilised the revenue will not cover the standing charges. A fifteen-minute service is, generally speaking, the lowest economic limit on an electric tramway. Every tramway manager tries to attract sufficient passengers for a more frequent service; and, as a matter of fact, it was found that where there was sufficient population the provision of a frequent and rapid service encouraged tramway travelling so much that cars had to be run at far shorter intervals than had been customary on horse tramways.

The increase of traffic brought with it the demand for larger as well as speedier cars with a shorter 'headway' or interval between one car and another. The capacity of a horse car is limited by the fact that it is not convenient to harness more than two horses to a single vehicle. But with electric cars there is no extraneous limitation to carrying capacity. Large double-decked cars with seats for seventy passengers are now quite common. In America it is a frequent practice to attach 'trailers' to the cars, making a short tramway train. Experiments have recently been arranged on similar lines in London, for the handling of the heavy traffic at rush hours. These instances show that electric tramway capacity is flexible and may be adjusted to the density and the fluctuating character of the demand.

Finally, it falls to be noted that the power consumed by a tramcar is, roughly, proportional to the useful work which the car performs. As already mentioned, it costs about as much to work a horse tramway when the cars are empty as when they are full, since the main item is the maintenance of the 'vital principle' of a certain number of horses independently of the traffic. But with electric traction the motors require less power when the cars are running light. And less current for the motors means less current generated at the power station—that is to say, less steam, less oil, less coal, less wear and tear. If more current is demanded, it is because more passengers are being carried and more revenue earned.

Reviewing the subject broadly, it is apparent that the adoption of electric traction on a tramway is not so much a step in advance as a beneficent revolution. The higher speeds with more frequent, more comfortable, and more commodious cars have created a volume of traffic far beyond what could have been handled with horse traction. The change also led to a great increase in the length of tramway routes and to the construction of new tramway systems. In 1898, when the electric tramway movement began in earnest, there were 1064 miles of tramway in the United Kingdom. Now there are 2562 miles, and the number of tramway passengers is more than double the total of third class passengers on the whole system of British railways. The number of tramway passengers carried during 1909-10 (the last period covered by the published official returns) was equal to about 62 times the estimated population of the United Kingdom.

While the traffic has multiplied in this remarkable fashion, there has been a heavy reduction in the fares charged. This has been made possible by the economical features of electric traction. In the old days a horse tramway had to spend about £80 to earn £100; an electric tramway need spend only about £60. With this reduction in the proportion of expenses to receipts, and with the greater volume of business, it became feasible to stimulate traffic still further by giving passengers much longer distances for their money. In fact, electric traction proved so economical that people began to imagine that there was no limit to the reductions which might be made with financial safety. However, there is plenty of evidence that a limit exists. In many cases it has been touched, if not passed, but the public continues to clamour for all sorts of concessions. These demands are a great compliment to electric traction, but they are a decided embarrassment to the tramway manager who believes in a reasonable margin between his total expenses and his total revenue.

CHAPTER V

THE MECHANISM OF AN ELECTRIC TRAMCAR: THE OVERHEAD SYSTEM

A rough idea has already been given of the elementary mechanism of electric traction—the combination of generating station, of cars fitted with electric motors, and of a sliding contact between the two. It is in connection with the sliding contact that the ingenuity of tramway engineers has been mainly exercised. Three distinct solutions were evolved for tramway work, giving rise to three systems—(1) the overhead or trolley system; (2) the conduit system; and (3) the surface-contact system.

The first system is now almost universal in the United Kingdom. Part of the London system is equipped on the conduit system; and the tramways at Lincoln and Wolverhampton are constructed on the surface-contact system. Beyond these cases the trolley holds the field. In the United States and on the Continent there is a larger proportion of conduit work, but from a practical point of view it would hardly be necessary to mention either conduit or surface-contact if it were not for the great engineering interest which they possess and for the controversies to which they have given rise.

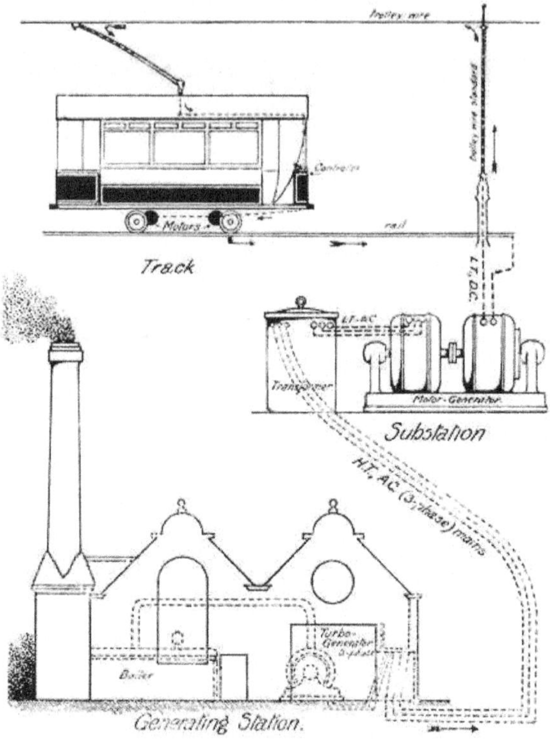

Fig. 3. Diagrammatic illustration of the general arrangement of an electric tramway on the overhead system. At the foot is shown the generating station which supplies alternating current at high-pressure (for economy in transmission) to a sub-station where it is 'transformed' to low pressure and 'converted' in a motor-generator to continuous current for distribution to the trolley wire from which each car takes its current. The course of the current through the trolley pole and controller and thence to the motors and back by the rails is indicated by arrows.

The overhead system has conquered because it is cheapest in first cost, cheapest to maintain, most economical in current, and most reliable in action. Later developments in surface-contact traction have run it very close on some of these points, but have not—for reasons which will be explained—affected the established position of the overhead system.

In its essential features the overhead system has not altered very much from the experimental line erected at the Paris Exhibition of 1881. The slotted tube has been replaced by a solid copper wire; and the 'boat' sliding within it has been replaced by a wheel or a bow pressed against the lower side of the wire by means of a pivoted arm controlled by springs. The sliding bow is common on the Continent, but it has been adopted on only one British tramway—that at Sheerness. Its use for electric traction on railways will be mentioned later, but as far as British tramways are concerned the bow is the exception which proves the trolley wheel rule.

The function of the trolley wheel is to collect current from the wire along which it rolls. This current passes through insulated wires down the trolley arm to the controller, which the driver of the car operates by means of a handle. The controller, which is really a series of electrical resistances, is analogous to a water tap. By its means the current may be completely shut off from the motors, or allowed to flow in varying degree as required by the speed of the car. In starting a car, the driver moves the controller handle notch by notch, so as to get a uniform rise in speed until the full current is allowed to pass through the motors. With such a mechanism, supplemented by brakes, the driver has the movements of the car under control.

In a four-wheeled car, each axle is driven by a motor. In a bogie car (one with a set of four wheels at each end) the axles of the larger wheels of the bogie are each driven by a motor; but not directly. Considerations of space make it necessary to keep the motor as small as possible, but if a motor is to be small and also powerful it must rotate at a high speed. On the tramcar, therefore, the motor drives a small toothed wheel which drives a large toothed wheel fixed to the axle, thus effecting a reduction of speed between the motor and the wheel.

The same considerations of space join with others in making two motors on each car the general rule. And the use of two motors enabled the tramway engineer to introduce a refinement into the method of control. This refinement is known as the 'series-parallel system.' One of its objects is to give a large 'starting torque' and so enable the car to gain speed quickly. When the current is first switched on by the controller it passes through the motors in tandem or in 'series,' thus dividing the pressure of the current (analogous to a 'head' of water) between them. The starting torque of a tramway motor (or the turning moment which it exerts when current is first passed through it) is dependent on the current but independent of the pressure. Thus the tandem or 'series' arrangement, which passes the full current through each motor, gives the maximum starting torque without an undue consumption of current. After the car is well started, the next movement of the controller puts the motors in 'parallel,' opening up two paths for the current instead of one, so that each motor receives the full pressure. The practical result is that there is a very rapid acceleration at starting, with marked economy in current. If the motors were kept in 'parallel' right through, twice as much current would be required to get the same starting torque. It will be seen later how valuable this arrangement for getting a rapid start, without excessive current consumption, may be in improving the physical and economic conditions of a tramway or train service.

After having passed through the motors and done its work, the current is led to the wheels of the car and returns by way of the rails, which are linked together by copper bonds so as to form a continuous conductor. The passage of the current from the wheel to the rail is indicated by sparks when the rails are rough or very dry and dirty. Although the rails, like the overhead wires, are thus carrying current, there is no danger of shock from them, as the electrical pressure in them is only a few volts, at the outside, while the pressure in the overhead wires is 500 volts. It is this difference of pressure which—like the 'head' of water in a turbine—supplies the motive power for the car.

Each car on a tramway system may thus be regarded as a bridge which completes an electrical circuit. When the driver moves his controller, current flows from the generating station at a high pressure, passes through the controller, operates the motors, and returns to the generating station at a low pressure. This typical circuit is completed through every car, so that the demand on the generating station at any moment is the sum of the demands of the cars at that moment. The business of the engineer at the generating station is to maintain the electrical pressure in the overhead wire at the normal level of 500 volts; and in order to do this on an ordinary tramway system it is found convenient to divide the overhead wire into half-mile sections, each of which has a separate main or 'feeder' from the generating station. The passenger can detect the change from one section to another by the click of the trolley wheel across the gap which insulates one half-mile section from another. At the same spot he can see the short square 'feeder-pillar' at the roadside (containing the switches by which current can be turned off from that section) and the cables which pass along the arm of the trolley standard and terminate in the overhead wire.

On an extensive tramway system the power-supply arrangements become more complicated. The central generating station remains the primary source of power, but sub-stations are erected at convenient points between the central station and the outskirts of the tramway area. These sub-stations are secondary stations for the distribution of electricity. They receive power at extra-high pressure (5000 volts or more) from the central station; they contain special machinery for reducing the pressure to 500 volts for distribution to the various tramway feeders. The object of this arrangement is partly technical but mainly economical. Electric power can be transmitted at a lower cost in mains and with less loss of energy at high pressures than at low. Consequently when the termini of tramway routes are several miles from the generating centre, greater all-round efficiency is secured by transmitting current at high pressure to a number of well selected sub-stations.

Fig. 4. Photograph of a car on a conduit section of the London County Council tramways. The centre line on the vacant track indicates the slot rail through which the 'plough' on the car passes to make contact with the conductors in the underground conduit. (Photograph reproduced by courtesy of Dick, Kerr and Company, Ltd.)

CHAPTER VI

CONDUIT AND SURFACE-CONTACT TRAMWAY SYSTEMS

Roughly speaking, the arrangements for generating electricity, distributing it, and utilising it on the car, remain the same in conduit tramways and surface-contact tramways as on the overhead system. The differences between the three systems are, as already indicated, confined to the means of collecting the current for each car.

Both the conduit and the surface-contact system were suggested as a means of escape from the main objection to the overhead system—the exposure of 'live' wires in the street. The cable tramway, with its concrete trough and slot, gave an obvious hint. There would be no difficulty, apparently, in carrying wires on insulators in the trough or conduit, and utilising the slot for a 'plough' which would slide along inside the conduit, keeping contact with the wires, and so conveying the current to the car.

This was tried for the first time in Blackpool, where—in 1884—a length of conduit tramway was laid along the front street of the town. The conditions could hardly have been less favourable for the system, as the sea frequently washed over the roadway, flooding the conduit with water and sand. Further, the conduit was so shallow that children were able to get at the conductors with their metal spades. As the conduit carried the return wire, the effect of a metallic contact between the two conductors was to cause a 'short circuit,' with very entertaining fireworks but with no amusing results for the tramway engineer. After a heroic trial, the system had to be abandoned.

Bournemouth was the next British town to adopt the conduit. It did so as a token of its exceptional civic pride. Three times, in fact, the Bournemouth Corporation declared that it did not want tramways of any kind whatever within its gates. And when the pressure of public opinion forced its consent, the arrangement was made that no overhead wires should appear in the central district of the town. Several miles of conduit tramway were therefore constructed (the trolley system being used for the outer tramway routes); and as by that time a good deal of experience had been gained in conduit work both in America and on the Continent, the contractors were able to give the Corporation a conduit system built to endure. At first the Corporation was reconciled to the fact that the conduit sections had cost about twice as much per mile as the trolley lines, but as years went on, and as the financial results of the system continued to prove unsatisfactory, the Corporation's contentment became modified. An examination of the accounts showed that the conduit sections could be reconstructed on the overhead system at a cost equal to the annual expense of maintaining these sections in good working order. Since the public had got used to the overhead wires on the other sections, and since they had not got used to owning tramways which produced a heavy loss, the decision was made to abandon the conduit system altogether.

In London the conduit system was adopted by the London County Council for various reasons. One was that the Council felt that London ought to have the best, the very best, and nothing but the best. Another was that the streets were so congested with traffic, lamp standards, telegraph and telephone poles, and other obstructions, that trolley wires and trolley standards would be a great nuisance and a serious danger. Aesthetic reasons were also advanced, but it is difficult to realise that they had much weight in connection with the majority of metropolitan streets. Trolley wires were, in fact, freely erected in suburban streets where there was a certain amount of beauty worth preserving.

The main underlying reason, no doubt, was the feeling that London could afford the most costly system. In any ordinary city (and perhaps in London as well) the conduit must be regarded as a luxury. It involves a continuous road excavation so deep that a great deal of incidental work has frequently to be done in moving gas, water, and drain pipes out of the way. The conduit itself is a thick channel of concrete, strengthened at intervals of a few feet with heavy cast iron 'yokes' which support the 'rails' forming the lips of the slot through which the 'plough' of the car passes. Elaborate arrangements have to be made for draining the conduit, as any accumulation of mud or water in contact with the conductors, or the special insulators supporting them, would be fatal to the working of the system. And in practice the ordinary drainage has to be assisted by continual scraping of the conduit with special brushes and by repeated flushing during the hours when the cars are not running. Heavy rains and snowstorms are therefore liable to upset the working of the system; and the tramway manager has to employ quite an army of men simply to keep the conduit in working order.

Trouble is also apt to be caused by purely mechanical means. On one occasion a child's hoop fell through the slot and caused a short circuit. As the ordinary scrapers slipped over the hoop, its presence was not detected for a considerable time, during which the tramway service was at a standstill. Altogether there is a greater liability to interruption on the conduit system than on the overhead system.

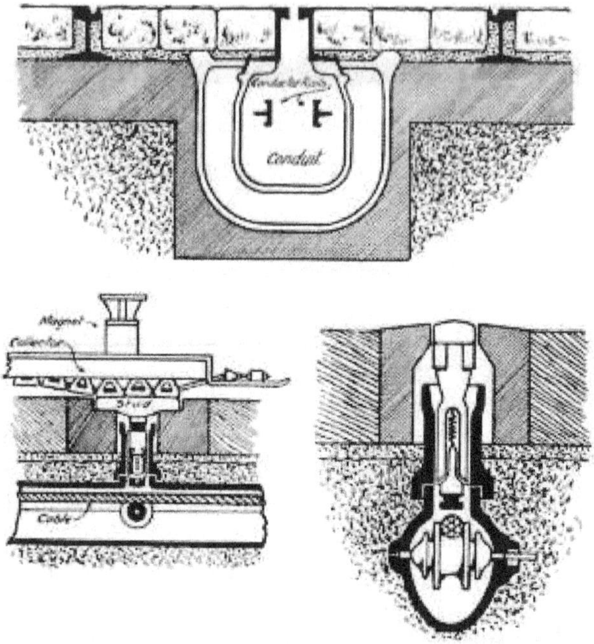

Fig. 5. The upper portion of the illustration shows a section of a typical conduit system of electric tramway traction. This section is taken at one of the cast-iron 'yokes' which support the rails forming the slot through which the 'plough' passes from the car to make contact with the conductor rails.

The lower illustration gives a longitudinal and transverse section of the 'G-B.' system of surface-contact tramway traction. The rope-like cable carries the current and is supported on insulators. When the collector on the car covers the stud, the action of the magnet draws the lower part of the stud into contact with the cable, thus supplying current to the car. After the car has passed, the lower part of the stud rises by the action of a spring and, breaking contact with the 'live' cable, becomes dead. (In actual practice contact would be made under the conditions shown in the left-hand diagram.)

Experience of these drawbacks led the London County Council to seek an alternative to the conduit when constructing electric lines in the north of London. Many of the borough councils, following the County Council's own previous arguments, would not listen to the suggestion of the overhead system; and a freshly-elected Council, pledged to a policy of economy, determined to try the surface-contact system. How this trial gave rise to a violent political controversy, leading to the abandonment of the project and culminating in important libel actions, forms a picturesque story which need not be told in detail here. Its main interest lies, for the moment, in the emphasis which the incidents give to a characteristic of the surface-contact system—its sensitiveness to minute alterations in detail.

The surface-contact or 'stud' system is really a modification of the conduit system. It has, in fact, been called the 'closed conduit.' The electric wires are again placed in a channel or pipe underground, but instead of being accessible through a slot, contact can be made with them only through metal studs placed at intervals flush with the roadway. By special electro-mechanical devices in the stud and on the car, the stud is brought into contact with the 'live' underground wire only when the car is over it. That is to say, the studs covered and protected by the car will be 'live' and supplying power to the car through a sliding brush or 'skate,' while those not so protected will be 'dead' and therefore of no danger to the public.

An immense amount of ingenuity has been expended by many engineers in devising studs to act with absolute certainty under all conditions. In the laboratory or the workshop, and even on an experimental track, it was simple enough to arrange a mechanism which would 'make' and 'break' contact with admirable regularity. But when it came to putting the mechanism down on an ordinary roadway, to be covered with mud, pounded by heavy traffic, and subjected to the action of damp, frost, heat, and all sorts of unexpected influences, much less satisfactory results were obtained. Time and again the hopes of engineers were dashed by a succession of petty troubles—some of them obscure, most of them unforeseen. The weak points in nearly all the systems were the insulation of electrical parts and the road construction work. Lack of simplicity and rigidity led to the introduction of moisture and to the shifting of parts so that studs jammed and remained 'alive' after the car had passed over them. But even after the practical elimination of these troubles the success of the surface-contact system seemed as sensitive as the system itself.

One system was tried at Torquay, and discontinued after a protracted trial on a large scale. Another system—the Lorain system—was installed at Wolverhampton and is still in operation, but without imitators. A third system—the Griffiths-Bedell or G-B. system—was installed in 1905 at Lincoln, with satisfactory results. It was the G-B. system which was offered to the metropolitan borough councils as an alternative to the conduit and the trolley. A trial section was laid down in 1898 in the Bow Road, and a certain amount of trouble was experienced with live studs and with various parts of the equipment. Owing to the stud system having been suggested by the Moderate Party, the experimental difficulties were extensively advertised by members of the Progressive Party, who condemned the system as dangerous and unworkable. Public feeling was worked up to such a pitch that, in the face of expert advice in favour of the system in a somewhat modified form, the Council decided to abandon the experiment. Libel actions by the owners of the 'G-B.' patents followed, part of the plaintiffs' case being that the system as laid down was altered in a number of small but vitally important details by the Council's officers and was therefore not the 'G-B.' system proper.

The results with the 'G-B.' system at Lincoln prove that it is possible to construct surface-contact tramways at a cost about 10 per cent. more than that of trolley tramways, and to operate them, safely and with reliability, at a cost not appreciably more than the general working expenses of an overhead line. But this proof has not only been enfeebled for the special reasons just described, but it came at a time when the public had got quite accustomed to the trolley and also when most towns had already been equipped with electric traction. Ten or fifteen years earlier, such a proof might have changed the course of tramway development; now it can have no great material effect.

The upshot of the contest between the three systems has, therefore, been the survival of the one which was most despised at the outset.

CHAPTER VII

THE BACKWARDNESS OF ELECTRIC TRACTION IN GREAT BRITAIN

Popular objections to the overhead system are not, of course, quite dead. Every tramway proposal in districts where the trolley has not already penetrated is still opposed on the ground of disfigurement and danger. This opposition serves as an index to the severity of the struggle which the advocates of the trolley system had to encounter before they made it almost universal in large cities. But the dislike of the public for a questionable novelty was not the sole reason why electric tramway enterprise was backward in Great Britain.

It is not strictly accurate to say that electric tramway *enterprise* was backward. The enterprise was there, in spirit, but circumstances were very much against it. Tramway schemes are controlled by special legislation which was passed before electric traction was contemplated; and this legislation has not been amended in any material degree to suit the altered conditions brought about by the use of electricity.

The Tramways Act, 1870—which is the master Act of the situation—was framed at a time of reaction against public monopolies. Before that time, gas, water, railway, and other companies had been granted statutory powers in perpetuity; and when a local authority wanted to take the supply of gas or water into its own hands, it had to buy the existing undertakings at the valuation put upon them by the owners themselves. There were frequent complaints about excessive purchase terms, and also about extortionate rates charged by the monopolist companies. Consequently, when horse tramways came on the scene, the legislature determined to put the new 'monopoly' on quite a different basis. The Tramways Act provided, first, that no application for tramway powers would be so much as considered if it did not gain the consent of the local authorities interested; second, that the period of tenure should be limited to twenty-one years; and third, that the local authorities should have the option, at the end of the period or at seven-year intervals afterwards, of buying the tramway undertaking at the 'then value' of the plant (rails, horses, cars, depots, etc.) without any allowance for compulsory purchase, goodwill, future profits or any other consideration whatsoever.

This Act was passed with the very best of intentions. It had the advantage of substituting, for the costly and clumsy procedure by Private Bill, the simple and cheap process of applying to the Board of Trade for a 'Provisional Order' which would acquire the full force of an Act when ratified (in a more or less automatic way) by Parliament. But in spite of its good intentions it proved a serious stumbling-block, especially when electric traction was proposed.

The effect of the limited tenure system, with compulsory expropriation on what were called 'scrap-iron' terms, was to make the companies very reluctant to spend one penny more than was absolutely necessary during the concluding years. Capital expenditure on improvements in equipment was regarded as out of the question, since there was not sufficient time to recoup the difference between first cost and the 'then value' at the purchase period. Money was grudged for the upkeep of track, the repair and painting of cars, and the hundred and one items of expense which are essential to a well-conducted tramway. System after system fell into a state of shabby gentility, hoarding money against its inevitable end.

This was the condition when, in the middle eighties, electric traction was suggested. The public, suffering from the decay of the tramway service, but not realising that the cause lay with an Act devised for the public benefit, expected the tramway companies to adopt the new mode of propulsion. But as the conversion to electric working involved track-work costing several thousands of pounds per mile, and new cars costing several hundreds each, together with a large generating plant and new car depots, the change was commercially impossible to companies which were forced to retain their old horse equipment in order to realise something for the shareholders in the day of expropriation. From these causes there arose a demand that the municipalities should take over the tramway systems and do what the companies appeared too slow to undertake.

Thus a strong impetus was given to municipal tramway enterprise. But this impetus did not remove the causes of delay. The local authorities had good economic reasons for waiting until the existing tramway leases ran out and so enabled purchase to be made upon the most advantageous terms. They were also obliged to move very cautiously in adopting so radical and so novel a change as electric traction. Municipalities are not speculative traders, who are ready to take risks after a rapid expert investigation of a new policy. Further, no municipality likes to accept the decision of another as valid for its own district.

The consequence was that each municipality thought it necessary to get its own expert report on the subject and, in many cases, to send its own deputation to inspect Continental tramway systems. These preliminary studies, with debates in Council chambers and newspaper columns, with public meetings of encouragement or protest, and with the erection of experimental lines, took up so much time that little of a substantial nature was done until several years after engineers were ready and willing to carry out the conversion of large systems of horse tramways to electric working.

The municipalities, however, were not the only forces at work. Towards the year 1896, when a large number of tramway leases were running out, a considerable amount of business was done by private capital in buying up horse tramways with a view to conversion and also to extension far beyond the limits of the existing routes. The essential condition of the success of such enterprise was, of course, the renewal of the tenure of the tramways for at least another twenty-one years. Here—and in the accompanying applications for extensions of route—the true inwardness of the Tramways Act was shown. Everything was in the hands of the local authorities. They had only to withhold their consent, and nothing could be done. And this power of veto enabled them to drive any bargain they pleased with the promoters of tramway schemes.

Most electric tramway proposals covered the areas of several local authorities, so that negotiations had to be entered into with each in turn. The municipalities, being the guardians of the public interests, considered it their duty to impose the heaviest conditions which the promoters could be induced to accept, rather than abandon the enterprise. It was a case of Hobson's choice in every parish. In some instances direct payments for wayleaves were demanded. In others the promoters were forced to bear the cost of street widenings and other 'public improvements' which were not always necessary for tramway purposes. In nearly every town the fares and stages were determined by the local authority—on the strength of the veto, not on commercial principles. The cost of construction was frequently increased by onerous conditions regarding the standard of overhead wire and track work. Under the Tramways Act, tramway companies were compelled to maintain the roadway between the rails and also outside for a space of eighteen inches—a provision which was sensible enough when horses were used. But the condition was not only enforced within these statutory limits when the promoters were about to use a form of traction which spared the road surface; it was extended in numerous cases to an obligation to pave the entire roadway and to maintain it—often with expensive wood paving where macadam had previously been considered quite good enough for the traffic.

One effect of this state of affairs was delay. The preliminary negotiations with local authorities—the interviews with mayors, aldermen, councillors, town clerks, and borough surveyors, to say nothing of the 'frontagers' along the line of route—usually occupied far more time than the actual construction of the tramways. They were also much more troublesome, since it was within the power of a single local authority in a central position to 'hold up' a complete scheme, while most districts had strong local patriotism and wanted a municipal system to themselves. Very little is known by the general public of the anxiety, difficulty, and expense attending such negotiations with local bodies divided into parties or cliques and furnished with an absolute power of veto. Looking back on the history of electric traction, it really seems extraordinary that engineers and financiers had the patience to undertake this work and carry it through. Their reward, as will be seen, was not great in a pecuniary sense; and, as regards reputation, they are generally accused of being extravagant, avaricious, and wanting in enterprise.

The ultimate effect was that the actual cost of electric tramways exceeded the estimates prepared on the basis of Continental and American experience. The more prolonged and difficult the negotiations preliminary to a scheme became, the greater the expense. And the conditions imposed by local authorities as the price of their consent loaded the capital account of electric tramway undertakings with items which had no direct concern with the tramway. The Board of Trade assisted the increase in cost by prescribing a standard of construction which was higher than that allowed in other countries. The net result has been that while electric tramways were expected to cost about £9500 per mile, they have actually cost over £12,000 per mile.

The revenue side of the account has also been affected by the power of veto. A local authority has no hesitation in imposing low fares and long stages (with high wages and short hours for employees) upon a tramway company seeking its consent. The standard usually adopted is that of large urban systems with dense traffic, so that systems in scattered districts are often unfairly treated. In municipal systems themselves the fares are apt to be determined by the promises of councillors at election times rather than by the simple consideration of a fair price for improved traffic facilities. Workmen's fares, for instance, are a dead loss on practically every tramway system. Every now and again there is an agitation for halfpenny fares, for the extension of stages, for cheap rates for school children, for free transport for the blind, and so on. A leading municipal tramway manager once remarked that it was almost impossible for men in his position to resist the pressure for such concessions, especially at local election periods. The chairman of the Highways Committee of the London County Council recently stated that never a day passes without some appeal for concessions in tramway fares.

Most of the large urban systems are under municipal control, and therefore they have the rates in reserve, as well as the most favourable traffic conditions, to encourage them in giving the public more and more for less money. But the tramway companies, working for the greater part in less thickly populated areas, with no extraneous means of making up losses, are put in a difficult position when similar concessions are forced upon them. The upshot is that the average return on the capital of electric traction companies amounts to only 3·41 per cent. Better profits were, in fact, made in the horse tramway days; and the electric traction industry is a fine example of the way in which the enterprise of engineers and capitalists may bring little comfort to themselves but enormous benefit to the public, which shows its gratitude by asking for greater blessings at their expense.

CHAPTER VIII

ELECTRIC TRAMWAY STAGNATION. THE TROLLEY OMNIBUS

The revenue of a tramway is built up of pennies; and a minute increase in the average earnings per passenger will therefore have a large effect on the total receipts. For instance, it was calculated (in 1907) that an increase of one-tenth of a penny in the average fare on the sixty systems under the control of the British Electric Traction Company would mean an increase of over £200,000 in the revenue. Similarly, a fractional decrease in one of the operating expenses—say, the cost of electric current—might transform a shaky undertaking into a sound one. Tramway finance, in fact, is a question of infinitesimals.

So long as fares are determined by arbitrary conditions, little can be done to increase the revenue on an electric tramway system. Such matters as the weather and the extent of building operations have far more influence on tramway traffic than anything the tramway manager can do to assist it. Apart from the development of parcels traffic, his best opportunities lie in the skilful adjustment of the service to the varying needs of the public, so that the 'rush' hours find an adequate supply of cars, while the quieter hours find no 'waste car mileage' in the form of empty cars. He can also do a good deal in the way of inducing the drivers not to waste current. By putting an electricity meter on each car it is possible to check the current consumption and, by a system of bonuses, to encourage the economical driver. There are many other directions in which small financial leakages may be arrested, giving an aggregate saving which is well worth the trouble.

Fig. 6. Photograph of an electric trolley omnibus built by the Railless Electric Traction Company Ltd. in 1909 and operated at Hendon for experimental purposes. Later cars built by this company are of a lighter and simpler design, but the illustration shows clearly the arrangement of a double trolley for supplying current to a vehicle which 'steers' like an ordinary motor omnibus.

The fact remains, however, that on the whole the electric tramway business depends upon too narrow a margin between costs and receipts. The recognition of this fact, coupled with the legislative difficulties already described, led to the practical cessation of tramway development in Great Britain at a point far short of what was once expected. At one stage, no doubt, people were a little too enthusiastic about electric traction. They imagined that electric traction would create profitable traffic along the most deserted of side streets. Acting on that theory, municipalities constructed—or forced tramway companies to construct—lines along roads which could never supply enough traffic to justify the expenditure involved. The interest on capital and other standing charges for an electric tramway route are so substantial that a certain minimum of traffic density must exist before any profit at all can be earned.

However, after every allowance is made for such local excesses of enthusiasm, the under-developed condition of electric traction in Great Britain remains conspicuous enough. A sensible relaxation of legislative restrictions would go a long way to improve matters—if, that is to say, financiers could be induced to re-enter a field in which they have had many disappointments.

Great hopes of improvement were entertained when the Light Railways Act, 1896, was passed. The primary object of this Act was to encourage the building of cheap railways for agricultural and fishery purposes, but it was drafted on lines broad enough to include electric tramways. Arrangements were made for State and local contributions to the cost of such schemes, in cases where subsidies appeared to be justifiable. The procedure in obtaining powers was made as simple and as economical as possible. Applications for 'Light Railway Orders' had to be made to the Light Railway Commission, one of whose members then arranged to hold a local inquiry into the proposal. If sanctioned, the scheme was passed on to the Board of Trade for approval, and the Order, if confirmed, thus secured the validity of a Private Act of Parliament.

Nothing was said in this Act about the consent of local authorities, or about limited tenure, or about expropriation upon scrap-iron terms. But the Light Railway Commissioners chose to interpret the Act in terms of the Tramways Act, with the result that, when there was any opposition on the part of local authorities, the tramway promoter using the Light Railways Act was not much better off than before. He had to face a new difficulty in a clause of the Light Railways Act, which provided that when the proposed light railway was of sufficient magnitude and in such a position that it offered competition with an existing railway, the scheme should be submitted to Parliament as a Private Bill—that is to say, should face the most costly and cumbersome procedure of all.

The Light Railways Act thus proved a great disappointment. Its failure to afford relief seems to have taken away the tramway promoter's last hope of genuine legislative betterment. He has resigned himself to things as they are; and the utmost he does is to assert, when occasion offers, that there are many districts which might enjoy the benefits of electric traction if means were provided for bringing every scheme directly before an independent tribunal for consideration on its merits alone; if arrangements were made for obtaining wayleaves and land on favourable terms, and if he were allowed to construct and equip the line on a less costly basis than the Board of Trade now demands, even in rural districts.

Pending that revolution, tramway authorities are seeking to develop a cheaper means of electric traction than the tramway. At the present stage, urban tramways have spread through suburbs towards villages and small towns which are anxious for better transport facilities but have not sufficient population to justify a tramway extension. Inter-urban tramway systems—those connecting towns with a network of lines—are also adjacent to such minor centres of traffic. From time to time attempts have been made to meet the demand by means of petrol omnibuses, but they have rarely been successful—partly, no doubt, owing to the difficulty of working a limited petrol omnibus service economically at the extremities of an electric tramway system.

The latest solution of the problem is the 'trackless trolley' or, more correctly, the 'trolley omnibus.' In the 1911 session over a dozen tramway authorities applied for powers to use this device; and, if the financial results of the first attempts are successful, there will probably be a considerable growth in this type of electric traction.

The trolley omnibus is a hybrid between the trolley tramcar and the omnibus. It is akin to the first, because it derives its power from an overhead wire through a flexible trolley pole. It is akin to the second, because it does not run on rails but is fitted with solid rubber tyres and uses the surface of the road in the usual way.

Roughly speaking, its electrical equipment is similar to that of a tramcar. The trolley pole conveys the electric current to the controller, which admits it to motors geared on to the back axles. There are, however, one or two important differences. The absence of a rail which might act as a return conductor necessitates the provision of a second overhead wire and a second trolley-pole to connect with it. Thus the electrical circuit is from the power station, along the first overhead wire, down the first trolley-pole, through the controller and motors, up the second trolley-pole, and back by the second overhead wire to the power station. Owing to the vehicle being a steerable one, the trolley-poles have to be specially designed to give plenty of free play sideways. The vehicle itself is similar in appearance to a single-decked motor omnibus, and it runs on solid rubber tyres or spring wheels.

The first thing which strikes one about the trolley omnibus in comparison with the electric tramcar is the cheapness in first cost. All the expense of concrete foundations, heavy rails, and granite paving is avoided. On ordinary roads the overhead construction is much less costly, as a single line of poles supporting two wires is sufficient for the up and down services. Estimates show that the equipment of a mile of roadway on this system will cost only from one-fourth to one-third of the corresponding tramway system. Following on this economy there is the saving in the cost of maintenance and repairs—a serious item on the ordinary tramway. In actual working, the system has the advantage that the vehicles can steer past slow-going traffic, thus avoiding the delay caused on tramway systems through carts having to draw out, away from the track, when overtaken by cars. This steering or 'overtaking' power enables a trolley omnibus service to be maintained without obstruction on a narrow roadway which would be badly congested by tramcars running on a rigid track. When there is only one pair of wires, two trolley omnibuses may pass each other (whether going in the same or opposite directions) by the simple process of pulling down the trolley poles of one car and swinging them out of the way for a few seconds. On a single-line tramway it is necessary to provide loops at intervals for crossing purposes and also to arrange the service so that cars arrive at the loops simultaneously.

The other side of the picture is shown when we come to look into the costs of working.

No matter how good the road surface may be or how excellent the design of the wheel, the tractive effort required for a trolley omnibus must be relatively greater than that required for a tramcar. Nothing demands a lower tractive effort than a steel wheel running on a steel rail. Consequently the trolley omnibus takes more power per ton moved than the tramcar. When the road surface is wet or uneven, or muddy or loose, this difference is of course multiplied. Another addition to the working cost is produced by the tyres, which, if of rubber, may wear away at the rate of 1-1/2*d.* or 2*d.* per mile per vehicle. Owing to the uniform control of speed afforded by the electric system, there is less jerking at starting or stopping than is general with a petrol-driven omnibus; but in spite of that advantage, tyre wear on a trolley omnibus must remain an important item. Something must also be allowed for the effect of vibration upon the car body and electrical equipment—an effect which is of course much less pronounced when a vehicle runs on rails.

The balance between these advantages and disadvantages is not easy to strike, even on a general basis. And it varies so much under local conditions that tramway engineers debated a long time before they decided in certain cases to try the trolley omnibus in extending their traffic facilities. All they had to go upon was the experience gained on certain Continental routes, where trolley omnibuses have been running for several years. That experience encouraged the hope that trolley omnibuses might be a profitable means of developing traffic in conjunction with a tramway system, and along routes which would not provide sufficient business for a regular tramway.

The simultaneous adoption of the trolley omnibus on a number of tramway 'feeders' gave rise to an impression that tramway authorities had discovered the wheel-on-rail system to be less efficient than the tyre-on-road system. As a general proposition, nothing could be further from the truth. Tramway authorities have adopted the new system in certain cases where the possible traffic is comparatively small, not as a substitute for tramways, but as an alternative to self-propelled omnibuses. The carrying capacity of a trolley omnibus is about twenty, while that of a tramcar is frequently as high as seventy. The speed of a tramcar runs up to twenty miles an hour, while twelve miles an hour is as much as is comfortable (to say the least) with a vehicle running with solid tyres on an ordinary road.

Therefore, where large volumes of traffic have to be handled swiftly, the tramway will remain. But where a twenty-minute or half-hourly service of small vehicles is sufficient for the available passengers, a system which is much cheaper in first cost is clearly more suitable, even though it may not reach the standard of economy in working set by the large urban tramway. That is to say, the choice between the two systems depends entirely upon local circumstances.

Fig. 7. The 'auto-trolley' system of electric traction applied to the haulage of goods in a German quarry. (From *Electrical Industries.*)

As an emphasis upon this statement, it is significant that many tramway engineers regard the trolley omnibus merely as the forerunner of a tramway. For this reason they favour the adoption of the particular trolley omnibus system where the overhead equipment is adaptable with trifling changes to tramway purposes. They argue that, in the case of a village of a few thousand inhabitants, situated a mile or so beyond the terminus of a tramway route, a trolley omnibus service will not only be sufficient for the existing traffic, but will show whether the traffic is likely to increase (through the stimulation of building enterprise) up to the point where it would make the laying of rails worth while. When that point is reached, the rails will be laid and the trolley omnibus vehicles put on some other route which is at one and the same time a tramway 'feeder' and a tramway 'feeler.'

CHAPTER IX

REGENERATIVE CONTROL

Before going on to discuss the 'accumulator' or 'storage battery' system of electric traction, reference should be made to an invention which holds the germ of great economies in electric traction. This invention is known under the name of 'regenerative control.'

It has already been explained that the dynamo is reversible—that is to say, a dynamo may act as a motor, or a motor as a dynamo. This fact is usefully applied in braking tramcars. When a car has gained speed, its momentum represents a certain amount of stored energy. In stopping the car, this energy has to be absorbed or dissipated in some way or other. One method is to utilise the friction of brake blocks on the wheels, or of skids on the rails themselves. With the electric car, however, it is possible to absorb the energy by making it drive the motors as if they were dynamos. The moving car drives the wheels, which in turn drive the motors; and the current so generated may either be absorbed in electrical 'resistances' or led to electro-magnets which are so placed that they exercise a retarding pull on the rails. In any of these cases a car which is being stopped, or is being 'held back' by the brakes when going down-hill, is wasting power. It is clear, therefore, that a great deal of power could be saved if the current generated by the motors in retarding could be pumped back, as it were, into the electrical circuit.

This is the problem of 'regeneration' which has fascinated many electrical engineers. The practical difficulties underlying it are very great; and perhaps the only man to get within measurable distance of surmounting them was Mr J. S. Raworth, whose system of regenerative control was tried on a number of tramway systems and installed on the Rawstenstall tramways in 1909. It cannot be said with confidence that all the difficulties have been overcome; on the other hand, it would be rash to say that they are insurmountable. Mr Raworth, at any rate, retains his faith in ultimate victory; and the theoretical beauty of the system is so complete that it is bound to retain its fascination.

The practical result of regeneration is to eliminate the effect of hills. A regenerative car in descending a hill gives back to the generating station some of the excess energy required to take it up the hill. In the same way each car, in coming to a standstill, gives back a portion of the energy required to start it. A regenerative tramway may thus be represented, from the energy point of view, as one in which all the cars are running at normal speeds on level roads.

Incidentally the regenerative system gives a very perfect control of the speed of the car on all gradients, owing to the regeneration which begins automatically when the motors start 'coasting.' It is a power-saver and a brake in one; and its efficacy as a means of control is so great that, if its incidental drawbacks could be avoided, it would be worth adopting for this purpose alone, both on electric tramways and on electric railways.

CHAPTER X

ACCUMULATOR ELECTRIC TRACTION. THE ELECTRIC AUTOMOBILE

The use of the accumulator or storage battery in electric traction affords a very good example of how a means of propulsion may fail in one set of circumstances and contrive to succeed in another. Its history serves to remind us that the problem of cheap transport is really a group of problems, each one of which demands a particular solution.

The accumulator is a device for storing electrical energy in the form of chemical energy. Its action depends upon the effect of currents of electricity on lead plates in a bath of sulphuric acid. The passage of the current through the battery produces chemical changes which enable the battery to give out current when required. As the battery may remain 'charged' for several days, and may be discharged slowly or quickly, it provides a means of 'storing' electrical energy. In practice, and under favourable conditions, the efficiency of the storage battery is about 80 per cent. That is to say, there is a loss of about 20 per cent. in the process of conversion and re-conversion.

Fig. 8. A modern electric automobile.—The electric battery is placed under the front half of the car, and the motors drive the back axle through chains. (British Electric Automobile Co., Ltd.)

Great hopes were once entertained of accumulator traction on tramways. The storage battery offered a means of escape from all the difficulty and expense of carrying electric mains overhead or underground. By fitting each car with a storage battery, it could be made an independent self-contained locomotive, capable of running a certain number of miles until the battery was approaching exhaustion. By providing centres where the batteries could be re-charged—or, to save time, replaced by batteries previously charged—a continuous service could be maintained on a tramway system.

The advantages of accumulator traction, apart from the saving in first cost, are the absence of obstruction and danger from overhead wires, and of the risk of a general stoppage of the service when the current at the generating station fails from any accidental cause. When accumulators are used, the conversion of a horse tramway to an electric tramway becomes a very simple matter. All that is required is to erect a generating station and provide each car with a storage battery and electrical equipment. This equipment, it may be mentioned, is substantially the same as with ordinary electric cars. The current flows from the accumulator through the controller and the motors back to the accumulator.

Many trials were made with this system in the early days of electric traction, but there are no survivals. The failures were due in part to weaknesses in the batteries and to the difficulty of handling them with proper care under the rough and ready conditions of tramway service. The main cause, however, was the inherent drawback of all locomotive systems—the fact that the tractor has to haul its own dead weight in addition to the weight of the car and passengers. Lead being one of the heaviest of metals, this dead weight was a very serious item on accumulator tramcars. It proved to be a fatal item when the attempt was made to run large cars on heavy gradients. The rush of current demanded in starting such cars up-hill was in itself too severe a tax on the delicate structure of the batteries. In practice, moreover, the necessity of bringing each car back to the depot for re-charging, after a limited journey, proved very troublesome. The more extensive the system and the more frequent the service, the more troublesome this necessity became. Even the most enthusiastic advocate of the storage battery was at last forced to admit that it was not applicable to a system of transport, which demanded comparatively high speeds with large cars on all gradients and over a range of several miles from the centre of power.

After the admitted failure of accumulator tramways, the storage battery was for some time used only on river launches and small private vehicles. The conditions in both cases—and especially in the former—are very favourable to its operation. On a river launch the weight of the battery is not a serious item, as it serves to some extent in the place of ballast. Launches, moreover, are generally required for trips of a limited number of miles up and down the river from the boathouse or charging station of the owner. In contrast with the tramway, there is no demand for rapid acceleration at starting or for abnormal power at intervals. The batteries discharge slowly and fairly evenly, and are not subjected to serious vibration. The electrical equipment is extremely simple, as the motor is fixed on to the propeller shaft and operated by a controller on the deck close to the steering wheel.

However, if economy were the only consideration, it is doubtful whether the electric launch would have survived against the competition of steam and petrol launches. It has survived because the simplicity of the equipment, its silent running, and the absence of heat, smoke and fumes, make it the ideal thing for river work. The hire of an electric launch on the Thames costs more than that of a steam launch, but plenty of people are willing to pay the additional charge to avoid the drawbacks of steam propulsion on a small vessel.

Similar considerations underlie the extensive use of electric broughams in cities. Such vehicles are required only for travel within a restricted area and on streets where the gradients are seldom severe. Their carrying capacity is generally limited to two or four passengers, so that the batteries do not require to be unduly heavy. A maximum speed of 12 miles an hour is quite sufficient for city streets; and with careful treatment the batteries can be very economically used and will not deteriorate nearly so rapidly as they would under tramway conditions. Considerations of economy, on the other hand, do not weigh very heavily with the class of people who use private electric broughams. They are prepared to pay for the best available; and the electric brougham, with its noiselessness, its easy running, its absence of smell or other nuisance, is regarded as the ideal which other modes of city transport must do their best to approach.

In London a certain amount of business has been done for some years in hiring electric broughams for various periods on terms which include current, maintenance, garage facilities, driver's wages, and all other charges. The convenience of such an arrangement to the hirer need not be emphasised, since what is wanted in this case is a vehicle which is always ready at a telephone call. But the system has another important advantage, which bears upon the economic prospects of accumulator traction. By retaining the vehicles under its control the hiring company not only centralises the arrangements for storing and re-charging, but it is able to take care that the batteries are properly treated. Just as the success of the surface-contact system depends on minutiae of design, so the success of accumulator traction depends upon minutiae of treatment. Carelessness in driving the vehicles and in handling the batteries at the garage may transform a perfectly satisfactory mode of city transport into an extravagant nuisance. Consequently the success of this class of business depends upon an organisation which permits of constant supervision over every vehicle and every driver.

A good deal of ingenuity has been exercised upon the electrical equipment of broughams; and it is probable that further improvements will be made. In some cases the front axle is driven by the motor; in some cases the back axle. The earliest cars used toothed-wheel gearing in order to reduce the speed of the small fast-running motor. Improved types on this principle still exist, but there are some interesting forms in which the motors are placed right at the hub of the wheels and effect speed reduction and control by electrical means, without any intermediate gearing.

In addition to these improvements, the storage battery itself has made a distinct advance in design and construction. It is more efficient, more durable, and more reliable now than ever it was before. The closer attention given to its treatment tends in the same direction; and the result is that storage-battery makers and engineers have a very accurate knowledge of what the accumulator will do at a certain cost under certain conditions. The conditions being the variable factors in the problem, and being in large measure determinable by choice, it is rather remarkable that the engineers and financiers should have selected, at the outset, the very conditions which were least suited to the peculiarities of the accumulator.

The attempt to adapt battery traction to tramway work is a conspicuous case in point, but it is not perhaps so conspicuous in the public memory as the efforts to organise electric cab and electric omnibus services in London and elsewhere. These efforts have been made so often and failed so regularly that they have made it difficult to obtain capital for any form of electric battery propulsion.

The electric omnibus has many of the drawbacks of the storage-battery tramcar, but they are not so serious in the case of an urban service, adequately met by small cars running at moderate speeds on short routes with moderate gradients. It is possible that if recent metropolitan electric omnibus enterprises had been as happy in their finance as in their engineering, they would have succeeded well enough. But even in their engineering they had to meet great difficulties. They sought to protect themselves against excessive costs by entering into maintenance agreements with the makers of the batteries; and although the terms of these agreements were satisfactory enough, their validity depended on careful treatment of the batteries by the drivers of the cars—a matter which it is rather difficult to guarantee. Moreover, the number of omnibuses put on the road was so small that the garage costs and other standing charges were proportionally very heavy. With a larger fleet and with efficient organisation, much better results might have been achieved in spite of the inherent difficulties of the situation.

Although the electric cab has the advantage of being a smaller vehicle and therefore more adapted to economical propulsion by storage batteries, the conditions of the cab service are not at all favourable to the system. The essential feature of a cab is that it should be available anywhere, to go anywhere at a moment's notice. An accumulator-driven vehicle, on the other hand, is tied by an invisible cord to the charging station. Even if charging stations were multiplied enormously, the electric cab would have no real freedom of action, since several hours are required for the process of re-charging. We have only to compare the limitations of the electric cab with the freedom of the petrol cab (which can renew its supply of petrol in a minute or two at any motor depot) to realise that the roving commission is not at all suited to the former.

In 1899 a very bold effort was made to establish an electric cab service in London. To inaugurate the service a procession of the cabs was formed, but it excited more ridicule than serious interest. The clumsy appearance of the cabs was against them; and their behaviour was not satisfactory enough—as to speed and reliability—to overcome the first unfavourable impressions. They soon disappeared, to add another failure to the long list of disappointments in connection with accumulator traction.

The private electric automobile remains, however, because it has been organised under conditions which suit the peculiarities of the storage battery. Its survival, in conjunction with the failure of a similar means of transit for tramway, omnibus, and public cab services, has pointed to another direction in which the electric automobile should be a commercial possibility. That is, in connection with the local distribution of goods from large stores and other centres.

The United States have given a very distinct lead in this matter. In New York, Chicago, Washington, and other large cities the electric automobile for private use is highly developed and there is also an extensive service of electric vehicles ranging in size from a small parcels van to a large lorry capable of carrying loads up to several tons. No doubt the local cost of other means of transport has something to do with this American development, which has, moreover, been strongly supported by the companies which supply electricity to the public. But the fundamental reason lies in the special character of the service demanded.

The vans belonging to a large store all start from a certain point and return to it after journeys of limited range. Owing to the period occupied in loading up, and also to the pre-determined hours of most of the deliveries, there is no difficulty about affording the time required for re-charging the batteries, or in arranging each journey so that the vehicle returns before the batteries are exhausted. With a standardised fleet of vehicles, it is possible to remove the discharged batteries and replace them with charged ones in a few minutes. The whole arrangement, in fact, is like a private automobile garage, with the advantage that the probable demand can be forecast with a somewhat greater degree of certainty.

Steam and petrol-driven wagons run most economically on long steady journeys at fairly high speeds, and the electric automobile does not attempt to compete with them on these lines. But it offers competition within city limits for door-to-door delivery; and its prospects are particularly good for light parcel service, where the horse is still maintaining its position against the petrol vehicle. The advantages of the electric vehicle in neatness and noiselessness will certainly secure its success if the cost can be proved to be not appreciably greater than that of its rivals.

Apart from the necessity of careful organisation, the main essential of success in electric automobile work is a supply of cheap electricity. Owners of private electric launches have to pay anything from 8*d.* to 2*s.* 6*d.* per unit for re-charging their batteries, but these high prices are due to the intermittent character of the demand and also (in some cases) to the cost of providing machinery to supply current at special pressures for particular launches. An electric automobile garage, situated close to a public generating station and offering a larger and more regular demand, will of course obtain current much cheaper. And it is possible that arrangements may be made for supplying electricity to automobiles at a much lower rate even than that customary for general power demands. In the metropolitan borough of Marylebone, for instance, an electric garage may obtain current during the small hours of the night at 1/2*d.* per unit, which is half the standard rate for power purposes. This low price is offered because there is otherwise practically no demand at all for electricity during these hours. If, therefore, a garage arranges—and the arrangement is quite feasible—to charge its batteries overnight, the power bill may be divided by two.

The electric automobile has been used to some extent as a touring car, but although journeys up to 100 miles have been performed on a single charge, the time occupied in re-charging, and the difficulty of finding convenient charging stations, are fatal to any development in this field.

CHAPTER XI

PETROL-ELECTRIC VEHICLES AND MAIN MARINE PROPULSION BY ELECTRICITY

Between the petrol-driven vehicle and the electric automobile there is an interesting series of links provided by 'petrol-electric' systems.

At one end of the chain, electricity plays an important part in supplying power to drive the car. At the other end, electrical apparatus is introduced merely as a form of transmission gear between the petrol engine and the driving axle. The reason for attempting the petrol-electric combination will be most readily understood by considering the latter arrangement first.

The petrol engine is a high-speed engine, capable of working most satisfactorily when it runs at a uniform rate with a constant load. On the other hand, the speed of the driving axle of a car varies from a very much lower speed down to zero. It is therefore necessary, when driving a vehicle with a petrol engine, to arrange some forms of variable speed-reducing transmission gear between the engine and the driving axle. The problem is further complicated by the fact that the petrol engine is irreversible, has practically no 'starting torque,' and has a very slight overload capacity. It has to be started running 'light' and then switched on to a low gear which gives sufficient power to overcome the inertia of the car. As the speed of the car rises, there have to be successive changes of gear. These difficulties are, of course, accentuated when dealing with the heavy weight of an omnibus.

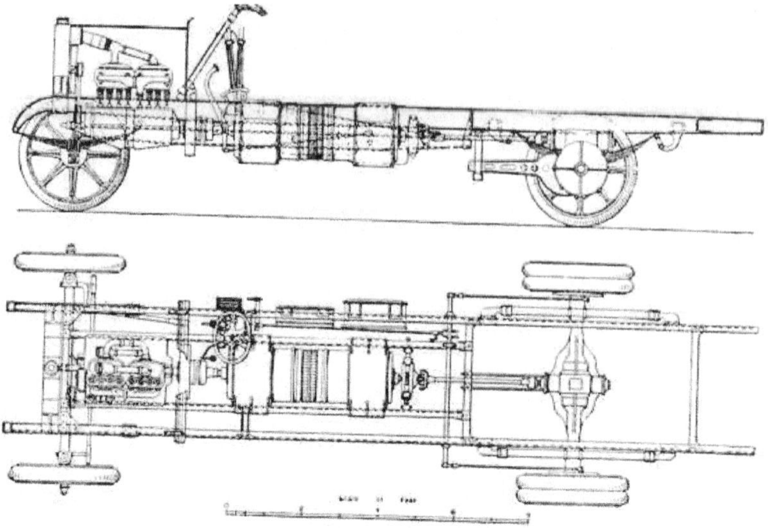

Fig. 9. Elevation and plan of a petrol-electric motor omnibus equipped by W. A. Stevens, Ltd. Directly behind the front wheels is the petrol engine, driving a dynamo through a flexible coupling. The dynamo supplies current to the motor directly behind it; and the motor drives the rear wheels through a cardan shaft. The transmission of power between the engine and the shaft is electrical at all speeds.

Practically all the troubles with petrol motor omnibuses have resided in the gear; and even the most ardent enthusiast for the all-electric faith must admit that the motor engineer has overcome these troubles (in great part if not wholly) with remarkable skill and ingenuity. But the complications of an adjustable mechanical bridge between a high-speed engine and a varying low-speed axle are so great that an electrical bridge was proposed as a substitute. By coupling the engine direct to a dynamo and by using the current so generated to drive variable-speed motors geared to the driving axle, the electrical engineer hoped to get better working results from the petrol motor than could be obtained with any mechanical transmission gear.

The most conspicuous advantage, apart from the quietness of running at all speeds, lies in the ease and smoothness with which the petrol-electric motor can start and gain speed. In this respect the combination system is practically on the same level as (or even superior to) the electric tramcar or the electric automobile. There is an entire absence of the jerks and jarring noises which usually accompany the starting of a motor omnibus. The same facility of control is of advantage in adjusting speed to suit the other traffic on the road, and also in negotiating hills.

In one class of petrol-electric vehicles the electric transmission gear is continuously used. In another, it is used at all speeds except the highest, when the engine is coupled directly (by a magnetic clutch) to a mechanical driving gear. In a third class the arrangement is more complicated, as it involves the use of storage batteries as an auxiliary to the power provided directly by the petrol engine. The Fischer type of petrol-electric vehicle uses electric transmission solely and has a fairly large battery to supplement the engine-produced current when steep hills are being negotiated. At ordinary speeds on level roads the surplus power produced by the engine goes to charge the battery.

The 'Automixte' type is peculiar in using the mechanical transmission gear all the time. The dynamo coupled to the engine supplies current to a small battery when surplus power is available; the same dynamo may be driven as a motor by current from the battery when such assistance is wanted at starting or on steep hills. The electric part of the equipment thus acts first as a generator and then as a motor, the change taking place automatically.

These different petrol-electric devices are very attractive from the engineering point of view, but at the present time it is uncertain whether they will realise the hopes of their inventors. The additional weight of the electric equipment is against them; and in some cases there appears to be a lower all-round efficiency. So that the motor-omnibus world, as a whole, continues to fix its faith upon the improved forms of mechanical transmission.

The underlying idea of the petrol-electric system has, however, been suggested for marine propulsion with a somewhat better prospect of success.

There is a partial analogy between the conditions of motor omnibus working and of ship propulsion with turbines. The steam turbine is, like the petrol engine, essentially a high-speed machine. The screw propeller, on the other hand, works most efficiently at low speeds. Therefore the marine engineer has to try and find some common denominator between an engine which runs most efficiently at high speeds and a propeller which is at its best when revolving comparatively slowly.

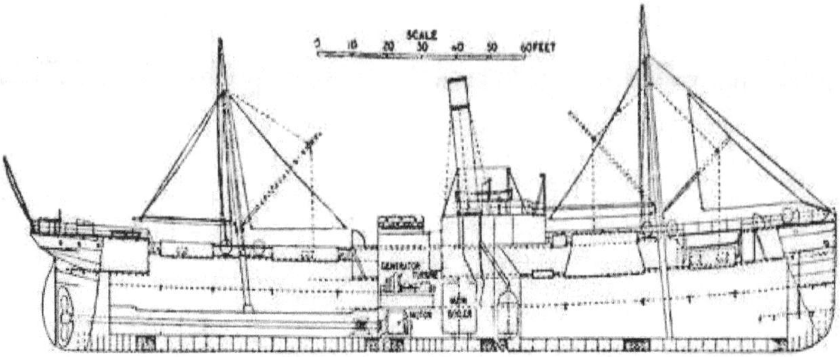

Fig. 10. Diagrammatic section of a steamship which has been 'converted' from the ordinary method of propulsion to the 'Paragon' system of electric main marine propulsion. The reciprocating engine has been replaced by a steam turbine, coupled direct to an electric generator which supplies current to a motor attached to the propeller shaft. The tests carried out with this vessel will indicate the advantages of the electric method of propulsion even with the usual long length of shaft. The vessel has a gross tonnage of 1241, and its speed is 9 knots. The engines replaced ran at 78 revolutions per minute and gave 500 brake horse power. The turbine now installed runs at 2500 r.p.m., and develops 630 brake horse power. (Illustration reproduced by courtesy of *The Electrician*.)

The gulf between the two has been narrowed by the improved design of propellers. Some engineers assert that continued improvements will bridge the gulf completely. Others have sought the solution in the same way as the motor engineer—by the use of mechanical change-speed gears. The suggestion has also been made to employ hydraulic gear as an intermediary; and in some recent vessels reciprocating engines with comparatively low-speed turbines driven by exhaust steam have been adopted.

In the electric system the turbine is coupled direct to an electric generator and may run continuously at the highest economical speed. The propeller shaft may be quite short and is driven by a slow speed motor connected by cables to the generator. Various arrangements for controlling the supply of current to the motor (with appropriate design of generator and motor) have been devised by Mr Durtnall, Mr Mavor, and other workers in this field; but whatever the details of these arrangements may be, they all give a wide range of speed both ahead and astern. The direct drive with the steam turbine has really only one speed—full speed ahead; and as the turbine is irreversible, 'astern' turbines have to be installed in addition. These limitations and complications are removed entirely when electrical transmission is adopted.

Moreover, the electric system can be so arranged that the control gear may be operated from the bridge itself. The facility in manoeuvring is, in fact, so marked that it would recommend electric marine propulsion even if that system offered no advantages on the score of economy in weight, space, and steam consumption over the existing systems. The steam turbine, it may be noted, has been adopted so far only in high-speed vessels; and it is generally recognised that its extension to vessels which run at 12 or 16 knots depends upon its adaptation to slow-speed propellers. Advocates of electric marine propulsion claim that they hold the most efficient solution of this problem.

It may also be pointed out that a considerable section of marine engineers look forward to the use of internal combustion engines (driven by oil or gas) on board ship. For naval purposes especially it would be a great advantage to do away with funnels and so leave the decks more free for gun mountings. As internal combustion engines are irreversible, the electric system offers a means of escape from a fundamental drawback to their use at sea. Here again the perfection of manoeuvring power, especially with twin screws (either of which may be controlled from the bridge through a wide range of speed ahead or astern), gives the electric system a strong claim for consideration by the naval authorities.

It is hardly necessary, except as a matter of curiosity, to refer to the suggestions made, from time to time, of accumulator-driven ocean steamships. Some wonderful pictures have been published of large vessels with tons of ballast in the form of storage batteries. They are likely to remain in this ideal condition, for although the driving of a large vessel by stored electricity is quite possible, it is also about the most expensive method which has ever been proposed.

Electric power from storage batteries has been used as an auxiliary in the propulsion and manoeuvring of submarines. In aerial navigation electricity has so far been employed to a very limited extent. Small airships have been designed to carry electric accumulators connected with various motor-driven propellers for raising, lowering, going ahead or astern, and steering. The switches which control the passage of the current to these propellers are connected with a wireless telegraph receiver, so that each operation may be started or stopped by a particular ether wave or series of waves. Demonstrations of such 'wireless-controlled' airships have been given in theatres; their field of usefulness, if any, is in connection with war on land or sea. Whether they will have any better fate than other devices for dropping bombs over the enemy's camps or ships remains to be seen.

One inventor has, I believe, suggested a means of direct electrical propulsion for aeroplanes, the current being derived from a petrol-driven generator and carried to motors attached to propellers so arranged as to give certain advantages in stability and manoeuvring. As yet, however, the probability of electricity being applied to locomotion in the air as well as on land and on sea is somewhat remote.

CHAPTER XII

THE PIONEER ELECTRIC RAILWAYS

Electric tramways have reached a period of middle age in which they are more concerned about their internal economy than the prospect of enterprise in new directions. Such development as they feel capable of making under present legislative conditions is only by proxy and tentatively, with the aid of the trolley omnibus.

Electric railways, however, have still many worlds to conquer. They are now in much the same position as electric tramways held about the year 1896. That is to say, they have already given practical proof of their capabilities and enabled engineers to point out the directions along which they are certain to develop. In the railway world there is a growing conviction that the adoption of electric traction on all suburban and inter-urban railways must be simply a matter of time. For main line traffic the possibilities of using electricity are as yet only an article of faith among electrical engineers.

Although the earliest experiments in electric traction were made in the railway form, the first electric lines could hardly be regarded as railways in the ordinary sense. They were really light railways, in which the traffic conditions approximated to those of tramways. The routes were short, the cars small, and the traffic of modest dimensions. They contained the germ of both the tramway and the railway; but, in the case of the railway, many years of technical development had to pass before the problem of applying electricity to the handling of large masses of traffic under standard railway conditions was solved.

The fact that the first electric railway in the United Kingdom was constructed at the Giant's Causeway (in 1883) is significant. The Giant's Causeway is one of the few places in our islands where water power is available close to a district with a demand for traffic facilities. In 1885 another electric railway deriving its energy from water-driven turbines was built between Bessbrook and Newry. At that period it was considered that waterfalls provided the only really feasible source of cheap electricity on a large scale. Even yet the impression survives that electric power stations using steam cannot produce current so cheaply as those which 'harness' waterfalls. Many people, in fact, are inclined to attribute the comparative backwardness of electrical development in Great Britain, not to legislative conditions, but to the lack of large waterfalls.

There might have been more active progress in the pioneering days if the presence of water power at convenient points had encouraged electrical engineers to repeat the experiments at Portrush and Bessbrook. But at an early stage in electrical history it became clear to engineers that coal was just as feasible a source of cheap power as water. The idea that a waterfall provides power 'for nothing' is one of those superficial conceptions which make the hardiest of fallacies. To 'harness' a waterfall requires a heavy expenditure of capital on conduits, pipe-lines, dams, and other works. The interest upon that capital is a heavy item, apart from the cost of maintenance and repairs. Waterfalls are situated in mountainous country, generally remote from the centres of industry; the water-power station, therefore, has to face the cost of transmission mains and the loss of energy involved in conveying the power to the place where it is wanted. Further, waterfalls and the adjacent ground belong either to individuals or to the State; and payment is generally exacted for the right to use them.

All these items have to be covered in the price charged for current to the public or to railway undertakings. Nature may provide the 'head' of water 'free,' but man has to spend money in utilising it, just as he has to do in mining and in obtaining heat from the coal which is also provided 'free.' Anything which is obtained 'for nothing' is generally worth nothing.

The full economies of generating electricity by steam power are not, however, realised until business is done on a large scale. As the first essential of a successful electric railway is a plentiful supply of cheap power, development from the experimental stage of Portrush had to wait until engineers mastered the art of producing electricity from large generators. They gained the necessary experience with electric tramways and in electric lighting. We have seen how, as regards tramways, legislation delayed and hampered progress. A similar cause was at work in connection with electric lighting. In 1882 an Act was passed regulating electric lighting on lines modelled upon the principles of the Tramways Act, 1870. Capitalists declined to work under this Act; and it was not until after 1888, when the Act was amended, that any money could be found in Great Britain for electric lighting schemes. This delay was a serious handicap not only to electric lighting but to the business of British electrical manufacturing, as there was, comparatively speaking, no demand for electrical plant for over six years. Meanwhile, matters had been advancing on normal lines in other countries; and when the demand came at last, the manufacturers on the Continent and in America were the only ones organised and ready to meet it.

These points must be touched upon in order to understand why so long a period elapsed between the pioneer electric railways and the real electric railway movement as we know it to-day. They also serve to explain the prominent part which American and German firms took in electrical developments here. Engineering and legislative conditions combined to retard electric railway enterprise so that it did not begin to take firm root in Great Britain until about 1890, and did not attain to any conspicuous growth until the beginning of the twentieth century.

Until after 1890 the only electric railways in Great Britain taking power from steam dynamos were those at Brighton Beach, Ryde Pier (Isle of Wight) and Southend Pier, opened in 1883, 1886 and 1890 respectively. These were all, of course, of short length. The Brighton Beach railway, designed and constructed by Mr Magnus Volk, was a unique piece of work. The rails were laid on heavy concrete blocks below high-water mark; and the cars were platforms raised on a light iron structure. Power was conveyed to the cars from wires hung on posts like the standards of a tramway on the trolley system. The unusual sensation of travelling over the water was enjoyed by hundreds of people until the difficulty of maintaining the track (owing to the erosive action of the waves) led to the railway being abandoned and another line of more ordinary character being laid on the level of the undercliff roadway.

The first indication of the genuine electric railway movement was given in 1893, when the Liverpool Overhead Railway was opened. This line was constructed to afford communication along the line of docks fringing the Mersey. The track was carried on a continuous bridge in order to avoid obstruction between the docks and the streets behind; and being overhead, there were serious disadvantages attached to the use of steam locomotives. Electric locomotives were therefore employed.

In this case, it should be noted, electricity was not adopted because it was more economical or efficient than steam. The reason lay with the peculiar situation of the railway. A similar reason decided the promoters of the City and South London Railway to try electric locomotives on their line. This railway, which was opened in 1890, was the first deep level or 'tube' railway in the world. Moreover, it was constructed and equipped throughout by British engineers, and at a time when the art of tunnelling was much less advanced than it is now. In the later and more imposing development of tube railways in London, the foresight and enterprise displayed by the pioneers of the City and South London Railway are apt to be overlooked. It was, however, the success of the original line from the Monument to Clapham which made it possible to raise capital for the Central London Railway (opened in 1900) and for the extensive tube railway system organised by the Underground Electric Railways Company of London.

On a deep-level railway, steam is, of course, out of the question. Even on the old 'Underground,' built close to the surface and furnished with frequent openings at the stations, and by means of ventilating shafts, the atmospheric conditions were abominable. The sulphurous fumes were indeed recommended for asthma and other complaints, but on a tube railway they would have been sufficient to cure every human ailment. Therefore the choice lay between electric traction and haulage by cables, compressed air, or some other innocuous system. Within these limits electricity was chosen on its merits.

The first railway in Great Britain to undertake conversion was one in which both the physical and economic troubles were exceptionally serious. The Mersey Railway is little more than a tunnel under the river, and it is distinguished by heavy gradients and by the continuous necessity of pumping out the water which drains into it. With steam traction the difficulty of ventilating the tunnel was an added trouble. Owing to these various causes the working expenses were abnormally heavy, and led ultimately to a receivership. Electric traction was adopted as the only possible cure. The pumping and ventilation arrangements were both reorganised for electric power; and the trains were equipped with electric traction on the 'multiple-unit' system, an arrangement—to be described in the next chapter—which is well suited to the economical handling of steep gradients. The practical result was a great increase in traffic, with a marked decrease in the proportion of expenses to receipts.

No other British railways, happily, were in so desperate a condition as the Mersey line, but all of them were, at the end of last century, feeling the effect of certain disquieting tendencies. These tendencies were most marked in connection with suburban and short-distance inter-urban traffic, which is quite distinct in character from the main-line traffic. We talk glibly enough of railway traffic as if it were a unity, but it is clear that very different considerations govern the traffic on a main line between, say, London and Glasgow, and those which control the traffic on London suburban routes or on a railway connecting the adjacent towns of the Potteries. Some railways have to deal with all three classes at the same time and occasionally on the same lines of rails. Electric traction has, so far, made itself felt only where the suburban or similar inter-urban traffic has been separable from the main line traffic.

The growth which took place in suburban traffic before and after the end of the century ought to have brought increased prosperity to the railway companies, but it did not always do so. Competition between the various companies led to a reduction in fares; Parliament, by establishing workmen's fares, forced the companies to carry an ever-increasing number of passengers at a loss, or at least without profit; wages tended to increase and hours of working to decrease—both affecting the cost of operation; rates and taxes became heavier and heavier with the growth of municipal expenditure; and a higher standard of comfort and efficiency was demanded by the public. In some instances the situation was aggravated by the competition of electric tramways along routes parallel to the railways. This competition was limited to point-to-point traffic, its maximum range being about three miles; but it was a grievance against which the railway companies protested very loudly, especially when the tramways were owned by local authorities to which the railways paid large sums in rates.

The general effect of all these factors was to reduce the margin of profit on which the railways were working. We have seen, in the case of tramways, how easy it is for a slight change in a frequently-recurring expense to have a serious effect in the aggregate. Railways are in much the same position; and the various influences at work upon the suburban traffic brought them face to face with the importance, if not the necessity, of finding some means of dealing with larger volumes of traffic on a basis more economical than that provided by steam locomotives.

This means they found in electric traction; but it may be noted that even railway engineers took some time to realise exactly what electric traction offered them. They were looking for something to reduce their annual expenses; and when they made calculations about electric traction they found that, when the expense of providing the electrical equipment was taken into account, the total cost of hauling the trains electrically on the existing schedule might be greater instead of less than the cost of steam haulage. They were therefore inclined to look upon the economic benefits of electric traction as an illusion.

In course of time, however, it came to be recognised that the function of electricity is not to act like a blue pencil on the debit side of the revenue account. Its essential purpose is to increase the volume of traffic. From the public point of view this is very much more valuable. Passengers are not directly concerned with means of reducing working expenses, but they are closely interested in the improvement of the frequency and speed of the service. The adoption of electricity on suburban lines has really been dictated by the demand for increased facilities. At the 'rush' hours of the morning and evening, when the great tide of workers flows and ebbs, the capacity of the steam lines was taxed to the utmost. And with the growth of population the difficulty of running sufficiently frequent trains became almost insuperable.

Apart from these particular necessities, the general features of railway economics point to the supreme advantage of increasing the volume of traffic in every possible way. In a railway, as in a tramway, the preponderating item is the cost of construction and maintenance; and unless a certain minimum of traffic is carried, the most economical working in the world will not secure a profit. The standing charges fall upon the idle hours as well as upon the busy; for every minute that a line of rails stands empty there is a loss of money. Railway progress depends upon reducing the proportion of idle hours; and that can only be done where there is scope for the growth of traffic, and where there is means—such as electric traction—of dealing with that growth on an economical basis.

In the succeeding chapter it is explained how electric traction enables a more frequent service to be run with advantage even on systems which were worked to the maximum limit possible under steam conditions. But in the meantime it will be interesting to trace the effect itself on a railway which soon followed the Mersey Railway in making the change from steam to electricity—the Metropolitan District Railway.

Fig. 11. An electric train on the Metropolitan District Railway, equipped by the British Thomson Houston Company. The front and rear cars and one intermediate car are equipped with electric motors, all controlled from the 'cab' at the end of the train. The controller handle may be seen close to the nearest window of the first car. The rail immediately in front of the foot of the guard is the conductor rail which conveys the current to the train. The rail between the track rails carries the return current.

Throughout the steam age the finance of the District Railway Company was as unattractive as the physical conditions of the railway itself. No dividend was ever paid on the ordinary shares; and even with the growth of London there was little prospect of any dividend ever being paid. When—about ten years ago—the late Mr C. T. Yerkes came over from America and obtained a controlling interest in the District Railway Company with a view to converting it to electric traction, he was regarded as a philanthropic enthusiast. Many of the shareholders themselves were reluctant to give their assent to the change; they preferred to bear the ills they knew than fly to others which might be introduced by an American financier.

But Mr Yerkes and those who worked with him had something more in view than the improvement of traffic on the District Railway. They acquired control of several tube railway schemes and obtained powers for new lines, so as to organise a comprehensive system of underground electric transport in London. They had sufficient faith in the traffic possibilities of London to find the enormous capital required to construct these tube railways and also to convert the whole District Railway to electric traction. The constructional work occupied several years; and after the lines were opened one by one, arrangements had to be developed for through-bookings among the various lines and between them all and the existing underground railways like the Central London Railway, the Metropolitan Railway (closely linked with the Metropolitan District) and the City and South London Railway. A systematic attempt was also made to develop the travelling habit in London by persistent advertising of the railway services and by increasing the frequency and rapidity of the trains. From these points of view the organisation of the network of lines comprehensively known by the title of 'Underground' is certainly unsurpassed.

The difficulties which had to be overcome in this great work were enormous, but there has been no break in the thread of progress. The 'tubes' are paying dividends which, though modest, are an encouragement to further developments. The finance of the District Railway has lost its element of chronic despair. Considered as a whole, the results prove that where there is the potentiality of large traffic, electricity is the instrument which must be applied. During the steam days, the most crowded part of the District Railway (the 'Inner Circle') carried a maximum of 16 trains per hour. With electric traction that figure has been raised to 40 trains per hour. And the remarkable thing is that with each increase in the service the traffic grows. Many people welcomed the electrification of the District as a measure of relief from the overcrowding on the steam trains during the busy hours. But with a service of trains more than doubled in frequency and also increased in capacity per train, overcrowding continues and the 'straphanger' has become an established institution.

It may be accepted as substantially proved that, on suburban and inter-urban railways in populous districts, electric traction is a means of increasing traffic and diminishing the proportion of working costs. Moreover, these results have been achieved in conjunction with substantial reductions in fares and with marked improvements in the comfort of travelling.

The engineering aspect of these changes has now to be considered.

CHAPTER XIII

ELECTRIC RAILWAYS FROM THE ENGINEERING POINT OF VIEW

When electric railways were first considered, the natural tendency of engineers was to follow the existing model and merely substitute electric locomotives for steam locomotives. In point of fact, however, the engineering method now adopted is an evolution from the tramway model, not from that of the typical railway.

A certain advantage was, of course, to be gained by replacing steam locomotives by electric ones. The greater 'starting torque' of the electric locomotive enables it to get a train up to full speed more quickly; and the capacity of the electric motor for taking heavy overloads assists the electric train in surmounting heavy gradients. Some advantage was also gained by producing all the power at a central source, instead of having a large number of steam locomotives, which are really power stations on wheels. But the electric locomotive had still to be made heavy enough to get sufficient grip of the rails; it had to haul its own dead weight; and it had to be made powerful enough to tackle a full-sized train on the steepest gradient with its complement of passengers, although the general demand upon it might be considerably less than that maximum.

The electric locomotive, in short, was an advance upon the steam locomotive, but it did not get past the essential drawbacks of the locomotive system. A locomotive is most economical when hauling full trains for long distances at a uniform speed; it is essentially a long-distance machine. The first demand for electrification came, however, from suburban railways, where the stations are close together and where, therefore, the speed is constantly varying from zero up to a maximum and back to zero again. The traffic also fluctuates between extreme limits; and there is obvious waste in having to run heavy locomotives and trains backwards and forwards during the slack hours. There was therefore a demand for some method of propulsion which would enable the length of trains and the consumption of power to be adjusted more closely to the variations in the traffic.

A step in the right direction was taken when the locomotive equipment was placed on a car, thus utilising the weight of the passengers to increase the adhesion on the rails. But the full advantages of electric traction were not realised until what is known as the 'multiple-unit' system was adopted.

The idea underlying this system is quite simple. If, instead of concentrating the motive power on a single locomotive or driving unit, we distribute it among the cars forming a train, we get the multiple-unit system. An electric tramcar and a trailer attached to another tramcar and trailer, with a third tramcar behind, would form a model for a multiple-unit train. By connecting the electrical equipments on the three tramcars—front, middle, and rear—it would be possible to control the train from either end or from the middle.

This is the principle upon which all the electric railways in Great Britain are now worked, with the exception of the City and South London Railway, where locomotives are still used and where the trains are comparatively short and light.

It will be seen that each multiple-unit train is readily divisible. A single motor car may be run, or a car with one or two trailers, or a long train made up of as many motor cars and trailers as the platforms will accommodate. And whether the trains are long or short, the power absorbed is in proportion to the length of the train and the load of passengers. By this simple means power is economised, and the railway engineer is able to reduce the proportion of idle rolling stock.

The adjustment of the length of trains to the fluctuations of the service is made easier by the absence, in the multiple-unit system, of the necessity of shunting at the termini. As a multiple-unit train can be controlled from either end, a more frequent as well as a more flexible service can be run. With steam traction the number of trains which may enter or leave a terminus is limited by the time occupied in shunting and by the necessity of leaving lines of rails free for that operation. With an electric train on the multiple-unit system, no more time is lost than the few seconds necessary for the driver to walk from the front of the train to the rear, which then becomes the 'front.' No lines have to be kept open for shunting locomotives, so that the available accommodation for trains is considerably increased. Some of the London railway companies have spent enormous sums in enlarging their terminal accommodation and have found that it is still inadequate to the demands of the 'rush' traffic. Electric traction therefore offers them an improvement of enormous value without the expenditure of a penny on station alterations.

The crowning advantage of electric traction lies, however, in the more rapid acceleration which it affords. We have already seen how important this item is on tramways. It is still more important on suburban railways, where a high average speed, in spite of frequent stops, is a vital matter.

On the District Railway the rate of acceleration in the old steam days was about 6 inches per second per second. It was, in fact, so low that the trains could not reach a fair speed before the brakes had to be applied to bring the train to a stop at the next station. With electric traction the rate of acceleration has risen to about 18 inches per second per second. On the Liverpool Overhead Railway a rate of 36 inches per second per second was reached in certain tests. Heavy starting currents are, of course, necessary to bring a train from rest to full speed at such a rapid rate, but it is quite possible for the electrical engineer, without being unduly extravagant in current, to accelerate a train more quickly than the passengers would find comfortable.

The practical result of rapid acceleration (combined with rapid braking) is not only to give a higher average speed but also to enable a more frequent service to be run. Owing to the block system on railways it is impossible for trains to follow each other closely in the manner of tramcars; and it is therefore of cardinal importance that no train should occupy a block for one second more than is necessary. Rapid acceleration becomes all the more important in this respect because of the difficulty of setting down and picking up passengers quickly. This difficulty is overcome in part by using saloon carriages with middle and end doors, in place of compartment carriages. At first the District Railway tried to help matters by operating these doors pneumatically, but the mechanism became unpopular after a number of late-comers had been pinched by closing doors. The management has reverted to hand operation; and it has probably achieved more by educating the public to move quickly than it would have gained with its too-perfect mechanical system.

London travellers have become so accustomed to entering and leaving trains quickly that it is possible for an observer to distinguish strangers by their slower movements on an underground railway. Thus the passenger, as well as the service, has been 'speeded-up.' The more frequent service of trains with a higher average speed would not have been possible, however, without an improvement upon the old methods of signalling. There is no need to dwell upon the weakness of the human element in railway signalling; and it will be clear even to the layman that the strain of handling traffic with a headway of one minute and a half, or less, would be more than men could stand. Automatic signalling had therefore to be adopted to obviate the risk of disaster.

Each train, as it leaves a block or section, 'clears' the signals for that block; and when any train attempts to enter a block against signals, the current is automatically switched off and the brakes applied. The system is so perfect that, in spite of the enormous traffic worked under it, there has been no failure and no accident. It is, of course, costly to install; and its cost can be justified (financially) only when the traffic is very heavy—that is to say, when the conditions make it almost a necessity.

The supply of electric power to electric railways is organised on practically the same lines as in the case of tramways. That is to say, current is generated at a central station, transmitted at high pressure to various sub-stations, and supplied from there at working pressure through 'feeders' to each section of the system. In the case of the 'Underground' system, most of the power is taken from a single huge electric station at Chelsea. Current from that station drives trains as far west as Wimbledon, Hounslow, and Ealing, as far north as Highgate and Golder's Green, and as far east as Barking.

This is a magnificent example of the concentration which gives economy. If each of the underground railways forming the system had erected its own generating station, the total initial outlay, on land, buildings, and machinery, would have been greater, and the cost of current would have been higher, owing to the smaller output and the more irregular demand which a single railway affords. The ideal electric power station is one which is constructed with the largest generating units and produces current at its maximum capacity throughout the twenty-four hours of each day. The Chelsea power station is nearer the ideal than a smaller one supplying a short railway could be. And a station of the latter class is, it may be noted, nearer the ideal than the arrangements on a steam railway, where the sources of power are scattered in hundreds of locomotives.

The concentration of power is therefore one of the many factors which have enabled electric railways to give a vastly improved service at lower fares.

With two exceptions—to be considered in the next chapter—the electric railways of Great Britain are constructed on the 'third-rail' system. They are thus a reversion to—or, rather, a survival of—the original type adopted by Siemens in 1879. The 'third-rail' is carried on insulators a few inches outside the track rail; and the motor cars are provided with a 'brush' or 'shoe' which slides along it and collects the current. In the centre of the track there is generally a second insulated rail to carry the return current, as it is more convenient, under railway conditions, to have a conductor independent of the track rails than to follow the tramway plan of using the rails 'bonded' together. In stations and at crossings the third or 'live' rail is protected by a wooden board in order to reduce the risk of shock to anyone falling on the line or walking upon it. The board is placed high enough over the rail to allow the shoe to pass freely.

As regards the motor equipment on the cars, tramway models have been followed very closely. The 'series-parallel' system of control is again adopted in order to get the high starting torque which gives rapid acceleration with moderate current consumption. The course of the current is again from the live rail, through the controller, through the motors, and thence to the return rail. The controller itself is more or less on the tramway principle; and the main modification in it is the arrangement which enables all the motors on a multiple-unit train to be operated by a single controller. This is done by connecting the controllers electrically and using electric power so that they all work in unison. Some companies use, for this purpose, compressed air controlled by electricity instead of electric power alone, but in both cases the principle is essentially the same.

Considered as a whole, the difference between a tramway and an electric railway on the third-rail system is a difference in degree, not in kind. The traffic is greater and the speeds higher, but both serve the purposes of comparatively short-distance transit. Indeed, within certain limits they compete with each other.

There remains to be considered another type of British electric railway which points the way to the extension of the new mode of traction to main line railways.

CHAPTER XIV

ELECTRIC TRACTION ON MAIN LINE RAILWAYS

On tramways, automobiles, and 'third-rail' lines, the electric current used belongs to the type described as 'continuous' or 'direct,' because the flow is always in the same direction. The other type of current is known as 'alternating,' as it flows backwards and forwards many times per second. There are several kinds of alternating current—single-phase, two-phase, three-phase, and polyphase—each produced from generators designed in a particular way.

It is not possible to give any adequate account of these different kinds of alternating current without going rather deeply into the theory of electricity. The ultimate practical point is that in transmitting alternating currents the circuits increase in number with the phases. Thus, three-phase current requires three wires, two-phase current three or four wires, and single-phase current a single circuit like that of continuous current[1].

Fig. 12. Photograph of a train on the electrified section of the London, Brighton and South Coast Railway. The overhead wire is suspended from cables stretched between insulators, and current is conveyed from it to the trains through a 'bow' which slides along its lower side. The photograph is taken from the rear part of the train. The front and rear cars are both equipped with electric motors.

Where current has to be conveyed economically over long distances, it is generally done in the form of alternating current at high pressure. For instance, the transmission from a tramway power station to the sub-stations is almost uniformly by three-phase current at, say, 5000 volts. When it reaches the sub-station, it is 'transformed' down to the working pressure of 500 volts and 'converted' from alternating to continuous current by means of rotary machinery. The transforming is done by a stationary piece of apparatus similar in principle to the familiar induction coil. An induction coil takes current at a few volts from a battery into its primary circuit and transforms it, by induction in the secondary circuit, into current of high enough voltage to give a long spark. A transformer can be designed to 'step-up' or 'step-down' the pressure according to the requirements of the case.

So much explanation is necessary to give some account of the alternating current railways on the Continent and thence of the single-phase system on the London, Brighton and South Coast Railway. The Morecambe and Heysham section of the Midland Railway is also equipped on the single-phase system.

Most of the earliest electric railways on the Continent derived their power from waterfalls and had to transmit it for a considerable distance. Three-phase current at high pressure being adopted for this purpose, the Continental engineers set to work to find some means of utilising the high-pressure three-phase current directly. They did this by carrying the three wires on poles alongside the railway track, and using three 'bow' collectors (in place of trolley wheels) to convey the current to transformers on the motor cars or locomotives. In these transformers the current was brought down to working pressure and then led to motors designed for three-phase current.

An immense amount of technical ingenuity was exercised in developing this system; and when the Metropolitan Railway decided to follow the District in electrifying its lines, a three-phase system was proposed. As the Metropolitan and Metropolitan District companies share the working of the Inner Circle, it was necessary that both should adopt the same system. The result was that the question between three-phase and continuous current working had to go to arbitration. After a long discussion of masses of technical evidence, Mr Lyttelton, the arbitrator, decided that the direct current system was better suited to the conditions of traffic on an underground railway in London.

The wisdom of that decision will not be questioned now. Three-phase motors do not give the rapid acceleration which is so urgently required on suburban lines; there are complications in speed control; and the necessity of having three overhead conductors is also a serious drawback. For comparatively long-distance traffic with few stops, however, the three-phase system is quite suitable. That is to say, it is a possible solution of the main line problem.

The great simplicity and flexibility of the power supply arrangements in the case of alternating current traction encouraged engineers to find something better adapted to ordinary railway conditions than the three-phase motor. Their problem was to find an arrangement which required one overhead conductor instead of three, and also provided a motor with the high starting torque and easy speed control of the continuous-current motor. After much theoretical and experimental work, they found it in the single-phase system, using a motor which is similar in many respects to the continuous-current motor but capable of being operated by alternating current.

On the advice of Mr Philip Dawson, the London, Brighton and South Coast Railway Company decided to experiment with this system on the double line connecting London Bridge and Victoria stations, about 9 miles long. Power is supplied to each track by a single overhead conductor carrying current at 6000 volts. Transformers are placed on the trains to bring the pressure down to 300 volts; the current is then led through controllers to single-phase motors in much the usual way. The reason for using so high a pressure on the overhead line is not only economy in transmission. If lower pressures were used, the heavy currents required for train propulsion would require a thicker conductor and correspondingly heavier supports. At 6000 volts it is possible for two double sliding bows to collect sufficient current for a heavy train from a wire which is comparable in thickness to the ordinary trolley wire of a tramway.

The power distribution arrangements, it will be noticed, are very much simpler than with continuous current on the third-rail system. There are no sub-stations with rotary machinery. Power is supplied direct from the generating station to the overhead line and is transformed down by stationary plant on the train itself. Single-phase traction represents, in fact, power transmission for railway purposes reduced to its simplest elements.

The overhead construction differs, however, in some important points from the tramway standard. The supports, which are in both bridge and bracket form, are stronger; the insulators are, owing to the much higher pressure employed, more massive; and a different means of suspension has been adopted. Each conductor is hung by links from two steel cables stretched chain-wise between the supports. This method of 'catenary suspension' enables the bow to slide along the wire without the jolts which are noticeable with a tramway trolley. Such smooth running keeps the bow continuously at an even pressure on the wire—an advantage which is of great importance at high speeds. The trains are arranged on the multiple-unit system.

The full financial results obtained on this railway have not so far been made public; but it is sufficient for our purpose to note that the Company, after more than a year's full trial, extended the system to the Crystal Palace and to Croydon. Further extensions are, it is understood, contemplated over the suburban lines to Sutton and elsewhere; and in course of time the conversion of the main line to Brighton will be undertaken.

Here we touch upon the most interesting aspect of this demonstration of electric traction on the single-phase system. The system was adopted in the first instance because the third-rail system would lead to complications and dangers which could not be permitted at crowded railway termini shared by all kinds of traffic, suburban and main line. But the advisers of the Company had also in view the possibility of development beyond the range of suburban traffic. They therefore sought a system which, while comparable to the third-rail continuous current in the handling of suburban business, would be adaptable to main line conditions, where infrequent stops and long runs at high speeds are the rule.

The adoption of electric traction on such a route as the Brighton main line would be a benefit in several ways. It would lead to a faster express service, as the high overload capacity of the electric motor enables it to take small account of gradients. It would also lead to a more frequent service, as the electric system is free from the conditions which force a steam railway to try to concentrate traffic on a limited number of long trains. Further, it would, by reducing the time lost in stopping and starting, bring the average speed of stopping trains much closer to that of express trains. All these improvements—assisted, probably, by lower fares—should lead to a great increase in the volume of traffic, thus reproducing the characteristic results of electric traction on suburban lines.

---

[1] An admirable explanation of alternating currents will be found in Mr Frank Broadbent's *Chats on Electricity*. (Werner Laurie, 1910.)

CHAPTER XV

CURIOSITIES OF ELECTRIC TRACTION

Like many other industries, electric traction has had its history brightened and made picturesque by curiosities of invention. Locomotion has, in fact, been a favourite field for the freak inventor; and some of his efforts with electric cars have been as weird and as fatuous as the most remarkable of perpetual motion devices.

One of these electrical monstrosities was, indeed, a kind of perpetual motion arrangement. It was invented about the year 1890 and consisted of a car equipped with accumulators which supplied power to a motor which drove a hydraulic pump, which in turn worked a dynamo supplying current to motors driving the axles of the car, and also to the accumulator for re-charging purposes. The inventor was so sure that he had got the better of the law of the conservation of energy that he provided his car with pointed ends, fitted with revolving fans to break down the air-pressure, in order that a speed of 125 miles per hour might be achieved. His name was Amen; and it provides a fitting comment upon his scheme.

Fig. 13. Illustration of Elberfeld-Barmen hanging electric railway. From *The Electrical Industry* (Books on Business), published by Messrs Methuen.

Several electric flying-machine ideas found their way on to the patent records. In 1893 a Frenchman registered a design for an air-ship with a cigar-shaped body and electrically-driven propellers. There was, however, more originality in an American idea that the progress of trains on the overhead railway might be assisted by the action of balloons in taking the weight of the cars off the rails. Curiously enough, other original inventors tried to get the opposite effect, by devising magnetic arrangements to increase the adhesion of the wheels to the rails.

More plausible forms of super-ingenuity have been exercised in connection with established modes of electric traction.

For the conduit system one inventor suggested a kind of reversion to the 'continuous valve' of the old atmospheric railway. The slot of the conduit was closed by a continuous series of springs which would be opened in succession by the plough as it passed along. This arrangement was actually tried on an experimental track in London. Another inventor proposed a novel plan for keeping the conductor in a conduit free from damp. The conductor was to be made hollow, so that hot air could be pumped through it to dry off any accumulated moisture.

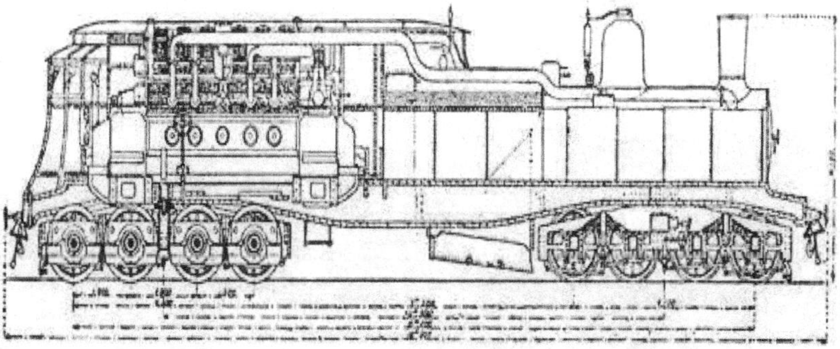

Fig. 14. The Heilmann electric locomotive—a generating station on wheels. The general arrangement of this locomotive should be compared with that of the modern electric turbo-locomotive described on p. 130 and illustrated in Fig. 15.

The most entertaining freak in connection with the trolley system was a device to enable two lines of car to use a single trolley wire. Cars going in one direction were to carry a double-ended inclined plane which would lift the trolley wheels of passing cars off the wire and let them slip back again. The only drawback to this arrangement was that it would not work.

Another inventor who was apparently impressed with the noise of trolley wheels on the wires designed a trolley head fitted with a pneumatic tyre. If he could have persuaded indiarubber to be anything but one of the best of insulators, he would have been completely successful.

One of the best known of electrical freaks—the Heilmann locomotive (Fig. 14)—is a very good example of the way in which an invention may be tried with enthusiasm, rejected with contumely, and revived at a much later date in an improved and more promising form. The Heilmann locomotive was practically a generating station on wheels. It carried a boiler and engines, which drove a dynamo, the current from which was led through controllers to motors coupled to the wheel axles. It was an enormous affair, over 18 metres long and running on sixteen wheels; extensive trials were made with it on the Western Railway of France in the early nineties. Some advantage was gained in smoothness of running, ease and uniformity of control, and improved acceleration; but its great weight, cost, and complexity were against it. In spite of the cordial support given to it by railway engineers, it was soon relegated to the scrap-heap.

Fig. 15. Electro-turbo-locomotive built by the North British Locomotive Company for experimental purposes. This locomotive is a 'generating station on wheels.' It carries a steam turbine driving a dynamo which supplies current through a controller to motors geared to the axles.

The Heilmann locomotive, it will be noticed, is similar in principle to the petrol-electric systems of propulsion now in use for road traction. But it is probable that the idea would never have been heard of again in connection with railway work had it not been for the appearance of the steam turbine. It was natural that the locomotive engineer should consider how the turbine could be applied to his purposes; and the first step in this inquiry made it plain that some electric method of control was necessary between the high-speed turbine and the driving axle.

Consequently, when the engineers of the North British Locomotive Company set to work in 1909 to design an 'electric turbo-locomotive,' they produced something not at all unlike the Heilmann locomotive. The equipment consists of a steam turbine, with elaborate condensing plant, a generator, and a group of driving motors (Fig. 15). The turbine runs at 3000 revolutions per minute and drives a continuous-current dynamo, the current from which passes through controllers to four motors which can be run in series, or two in series and two in parallel, or all in parallel, according to the draw-bar pull required. Trials with this locomotive were begun early in 1910, but it is yet too early to say whether it will be more fortunate than the Heilmann locomotive, and whether it is likely to delay the advance of the electric locomotive proper, fed with power by overhead wires from a central power station.

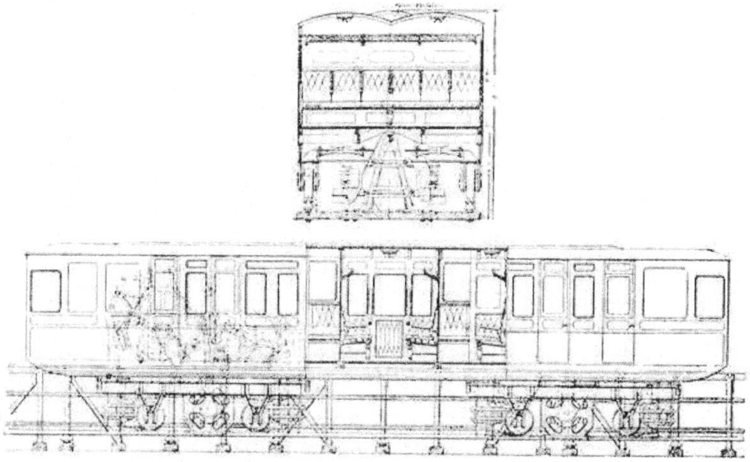

Fig. 16. Diagrammatic sections of the Behr electric mono-rail car. The car is balanced on the summit of a continuous trestle and is designed for speeds up to 120 miles per hour.

The possibilities of high speed on a mono-railway, and especially an electric mono-railway, have acted like a will-o'-the-wisp to the imaginations of many engineers. Of the various systems suggested, only one—the gyroscopic mono-railway invented by Mr Brennan—seems likely to survive; and even in that case victory under practical conditions is not yet certain.

At Ballybunnion there is a steam mono-railway which has been at work since 1888. It has had, so far as I am aware, no imitators; but its engineer, Mr Behr, retained so much faith in the principle that he decided to apply it to the problem of high-speed electric traction. During the 1900 session he promoted a Bill for the construction of a mono-railway between Liverpool and Manchester. There was tremendous opposition from the existing railway companies, which brought experts to prove that Mr Behr was a vain dreamer; but the Bill succeeded. The promoters, however, found it much harder work to raise capital for the project. They needed close upon £3,000,000, but the public response to the first invitation was so small that the scheme was abandoned.

The line, as projected, was nearly 35 miles long; and a speed of 100 miles per hour was intended, reducing the time of the Liverpool-Manchester journey to twenty minutes. At each end of the line (which was a double one) a steep gradient was arranged to facilitate starting and stopping—an arrangement, by the way, which is adopted to a certain extent on London tubes. The track itself was shaped like an inverted **V**, and practically the whole of the weight of the cars was borne upon a rail at the top. The wheels, therefore, were right in the centre of the car, which balanced itself on the trestle with its centre of gravity below the rail. Each side of the trestle carried two guide-rails which bore against free-running horizontal wheels on the car to prevent any undue lateral movement. Each car was designed to carry four motors with a total normal capacity of 160 horse power and an overload capacity up to 320 horse power. The rails for carrying the current were placed on the track in very much the same position as the ordinary rails occupy on a normal railway.

In another form of mono-railway—the Kearney high-speed railway—the wheels are placed below the car and run on a single rail laid direct on sleepers. The cars are held upright by flanged wheels on the top, running on a rail fixed to the roof of tunnels or to standards not unlike those of an overhead trolley. This railway has been exhibited in the form of a model.

Fig. 17. The Brennan gyroscopic mono-railway.—The car is electrically driven, and its equilibrium is maintained by the action of two gyroscopes, also electrically driven.

Mr Brennan's gyroscopic mono-railway was first shown, in a small size, at a conversazione of the Royal Society in 1907. Full-sized cars were constructed later, and one was seen at work during the Japan-British Exhibition of 1910. The distinguishing feature of the vehicle is the use of two gyroscopes (electrically driven), one horizontal and the other vertical, to maintain the car upright on a single rail, even when loaded unevenly and running at a fair speed round sharp curves. From one point of view, the gyroscopic car is no more wonderful than a spinning top, but the spectacle of a vehicle running steadily on a single rail was so extraordinary that the interest of the whole world was immediately aroused. Support was given to Mr Brennan's experiments by the India Office and the Colonial Office, on the ground that a railway which required only one rail, and was more or less independent of both curves and gradients, would be of great value in districts where the ordinary two-track railway might be both inconvenient and too costly. One drawback to the arrangement is the necessity of fitting each vehicle with gyroscopes, which are expensive and delicate pieces of apparatus. But the ingenuity of the invention is so great that Mr Brennan ought to reap the reward of seeing a gyroscopic railway in full operation before long.

Fig. 18. The 'Telpher' system of electrical locomotion adapted to the transport of materials in a factory. The 'car' is suspended from a girder and is operated by the driver in the same way as an electric car. (From *Electrics*.)

The only electric mono-railway actually at work is the 'hanging railway' at Elberfeld in Germany (Fig. 13). This railway is an evolution from the system of 'telpherage' which was devised in the very infancy of electric traction for the transport of goods. The root idea is to make the overhead wire carrying the current the track rail as well, the whole contrivance—rails and cars—being suspended from girders or cables supported by a series of standards or bridges. At Elberfeld the cars pass over streets and also over canals. There are no signs, however, that the 'hanging railway' will have any imitators. In appearance and in cost of construction and operation it does not seem to have any conspicuous advantages over a double-track overhead railway. The system of telpherage is therefore likely to be confined to the carriage of goods from one part of a factory to another, and (in the form of cable-ways) to the handling of materials in mines and other extensive engineering works. For such purposes it is having an increasingly extended application.

CHAPTER XVI

THE FUTURE

Nothing irritates an electrical engineer more readily than the repetition of the phrase, 'Electricity is in its infancy.' The words have been used by countless mayors and aldermen while 'inaugurating' tramway or electric lighting schemes; they have been echoed by innumerable journalists who persist in maintaining a Jules-Verne attitude towards the electrical industry. And what disturbs the electrical engineer is not only the banality of the phrase but the use of it as a comment upon the achievements to which he has devoted his life.

Nevertheless it will be admitted, from the rapid survey which we have taken of electric traction, that the potentialities of electricity in locomotion make an even stronger appeal than the actualities. Except in one field—the tramway field—engineers have only touched the fringe of possible developments in electric locomotion.

Even in tramway work we may, if legislative conditions improve and if current becomes much cheaper, see a considerable development in passenger and also in agricultural lines. Meanwhile the trolley omnibus offers a prospect of extension in electric road traction; and there is a great deal yet to be done with petrol-electric vehicles and with electric automobiles in certain classes of transport.

The great field, however, lies in railway traction. There are 200 miles of electric railway in the United Kingdom; and there are nearly 13,000 miles of steam railway. Not even the most sanguine electrical missionary will believe that this difference can be materially altered within the next decade, but there is ample ground for faith in the steady increase of the electrical figure. If the advance of electric traction on railways must be slow, it is because financial and not engineering considerations govern the speed of conversion. No railway company can take a step involving hundreds of thousands of pounds, and a revolution in working methods, without prolonged consideration and elaborate preparation.

On roads, on tramways, and on railroads, the future lies with electricity—wholly on railroads and tramways, perhaps not wholly on roads. There is scope for it also at sea; and if our canals are worth the cost of reconstruction on modern lines, electric haulage will be used there on the model of the canal haulage installations which exist here and there on the Continent. For marine work the advantages of electricity have yet to be confirmed by practical experience; but on land it has already proved that it supplies a means of locomotion which is more efficient, cleaner and more attractive, and more closely adapted to the needs and distribution of modern population than any other.

The fashion for devising Utopias is not so popular as it used to be, but in every ideal world which is more than a spiritual vision, and in every intelligent forecast of an advanced civilisation, universal electric transport is taken for granted. Electrical engineers are ready to prove that this standard element in Utopia is available at the present day on the basis which is the ultimate justification of all engineering projects in this workaday world—the basis of profit.

Their confidence will be intensified when we approach the 'all-electric' age prophesied by Mr Ferranti in his Presidential Address to the Institution of Electrical Engineers in 1910. Mr Ferranti looks forward to a national scheme for the supply and distribution of electric power. Under this scheme, the production of electricity would be concentrated in one hundred huge power stations, using engines of enormous capacity and acting as wholesale suppliers of electrical energy to towns, railways, tramways, and factories. The price of electricity would then be a fraction of what it is now; and all the economies of electricity in action would be multiplied accordingly. Technically, the scheme is quite feasible; and it could be realised in the near future if capitalists and the Government could be brought to appreciate the tremendous stimulus it would offer to industrial activity and the effect it would have in conserving the power which is latent in our coal measures.

Made in the USA
Monee, IL
08 November 2026

67946288R00052